THE DATA SHEPHERD:
Debugging the American Dream

ABDERRAHMAN A. El HADDI

The Data Shepherd: Debugging the American Dream

Copyright © 2026 by Abderrahman A. El Haddi

All rights reserved. No part of this publication may be reproduced, distributed, or transmitted in any form or by any means, including photocopying, recording, or other electronic or mechanical methods, without the prior written permission of the publisher, except in the case of brief quotations embodied in critical reviews and certain other non-commercial uses permitted by copyright law.

Data Shepherd Books Chaska, Minnesota

Publisher's Cataloging-in-Publication Data Names: El Haddi, Abderrahman A., author. Title: The Data Shepherd : debugging the American dream / Abderrahman A. El Haddi. Description: First edition. | Chaska, MN : Data Shepherd Books, 2026. Subjects: LCSH: El Haddi, Abderrahman A. | Shepherds—Morocco—Biography. | Computer scientists—United States—Biography. | Immigrants—United States—Biography. | BISAC: BIOGRAPHY & AUTOBIOGRAPHY / Personal Memoirs. | BIOGRAPHY & AUTOBIOGRAPHY / Science & Technology.

Paperback ISBN: 979-8-9945180-0-7

eBook ISBN: 979-8-9945180-1-4

First Edition: February 2026

Table of Contents

CHAPTER 1: Goodbye, My Goats 1

CHAPTER 2: Trouble in the House and the Will to Live 9

CHAPTER 3: A Busy and Scary Summer 19

CHAPTER 4: The Eviction and the Wedding 27

CHAPTER 5: The 1972 School Year 35

CHAPTER 6: Crime and the Angel 41

CHAPTER 7: Another Summer, Another Coup d'État 49

CHAPTER 8: The Boarding School and the Solidarity of a Good Friend 59

CHAPTER 9: The Holidays, the Executions, and a Rebellion 65

CHAPTER 10: In Memory of Dean Bekkali, Farmer Khammar, and Professor Pascon 73

CHAPTER 11: Goodbye France, Hello Minnesota 79

CHAPTER 12: Hello, University of Minnesota 87

CHAPTER 13: Goodbye, My Dear Friend 95

CHAPTER 14: The Experimental Station 103

CHAPTER 15: The Thesis Defense 109

CHAPTER 16: Three Jobs in a Month 115

CHAPTER 17: The Professor and the Wedding 123

CHAPTER 18: The Decision to Immigrate 129

CHAPTER 19: The New Immigrant 137

CHAPTER 20: The Porter's Price 143

CHAPTER 21: The Law Has Changed 147

CHAPTER 22: The Pivot to Land a Job 151

CHAPTER 23: Ali is Here 157

CHAPTER 24: The Government Grind 165

CHAPTER 25: The Bridge to Freedom 171

CHAPTER 26: The Dotcom Death March 179

CHAPTER 27: September 11 187

CHAPTER 28: Consolidated Data Systems Startup 191

CHAPTER 29: Outsourced Freedom 195

CHAPTER 30: Abdou's New Startup 199

CHAPTER 31: Goodbye Mother 203

CHAPTER 32: The Global Systems 207

CHAPTER 33: Goodbye Father 211

CHAPTER 34: The End 215

ACKNOWLEDGMENTS 217

ABOUT THE AUTHOR 219

CHAPTER 1: Goodbye, My Goats

Abdou was born in the small village of Timoulilt in Morocco to impoverished parents. Both were illiterate. His father worked as a school janitor and a cook, while his mother tended to his eight siblings and grandparents. They toiled in his aunt's fields, removing weeds and harvesting olives and almonds. In the summer, his father cooked at weddings and ceremonies in exchange for a few Dirhams a night, a skill he learned while working for the French before independence.

People treated his father as an outsider because he was not born in the village. Abdou despised this discrimination and swore he would raise the family status one day. Seeing his father work so many jobs just to make ends meet was painful.

Abdou had eleven cousins, ten boys and a girl, but he was closest to Ali. Ali was mischievous, courageous, and clever. Other kids feared him; he was a few years older than Abdou and built for trouble.

Abdou finished fifth grade on June 30th, 1970. He spent the summer with Ali, roaming the village's olive groves and moving their goats and sheep from one green spot to another along the waterway. They hunted doves and other small birds and swam in the crystal-clear springs that served as the village's only source of drinking water. No one in town had running water or electricity.

There was an unspoken rule in the village: no one swam or washed in the stream after five p.m., or they risked the wrath of the elders. Young boys, girls, and old women carried giant clay jugs on their backs or on donkeys to haul home clean drinking water.

Before sunset, Abdou would guide his donkey to the springs, load two giant clay jugs, and bring them home. Because Abdou was short and petite, his mother and grandmother had to help load and unload the twenty-five-liter vessels. He made the three-kilometer trip two to three times a day.

His grandmother, Itto, constantly reminded him to avoid the boulders lining the hilly, narrow path; he had broken jugs before, forcing the family to borrow from neighbors. Despite the labor, Abdou enjoyed the chore because he saw his shy classmate, Rkiya. They never talked—his heart hammered too hard whenever he saw her beautiful lips—but they exchanged discrete, forbidden smiles.

One day, Rkiya's older brother, a sixteen-year-old fourth-grade dropout, caught Abdou and the girl exchanging tender looks. He beat both of them severely. Abdou swore to take revenge. Rkiya, only eleven, was already promised to a forty-year-old teacher to be wed as soon as she turned twelve. She did not even have breasts yet. It was a case of legal pedophilia supported by religious fanaticism.

The teacher drove a nice green Renault 4. He was missing two bottom front teeth from excess tobacco and a cheap red wine called *Doumi Boulbbadre*. Abdou knew the brand well; he gathered scattered bottles and sold them to the corner store for a few cents each.

In the evening after the fight, Ali noticed Abdou's bloody left eye. "Who hit you, son of my aunt?"

"Rkiya's brother, cousin."

Unlike Abdou, who was short and chubby, Ali was slim. He had the biceps of a fighter and a deadly left hook. He was always the first to throw a punch in a skirmish, and he wasn't afraid to bite if challenged. The next day, Abdou went to fetch water, and Ali followed him like a wolf.

Rkiya was there again. Her face was covered with a red-and-green scarf to hide the marks her brother had left. It looked like her brother had gotten the best of her once she got home. Abdou felt a deep sadness.

Rkiya's brother appeared again. He glared at Abdou. "What are you looking at? Get your water and leave, *Dinmook a lakqlawi*!"

The curse—a blasphemy against his mother's faith coupled with a crude insult to his manhood—left Abdou steaming.

Before the bully could finish his vile insults, Ali jumped him. *Boom.* A series of punches and a few kicks to the groin dropped him.

"If you ever hit my cousin again, I will remove your pants in public, you *lakqlawi*," Ali roared, insulting the bully's manhood.

"I will tell your aunt!" the boy cried.

That evening, Rkiya's mother came to Abdou's home to complain. Abdou worried he had lost Rkiya forever. After supper, Abdou and Ali were punished with fresh, long olive sticks. The pain was unbearable. It was amazing how the olive tree could signify peace and punishment at the same time.

The two kids spent the night on the flat roof of the house to escape the heat. Sleeping on the roof was lovely; they watched the stars all night, though the noise from the stupid roosters kept them awake. As soon as light touched the sky, swarms of flies attacked their exposed heads and limbs.

Summer ended quickly. Abdou and Ali walked to the nearby village of Afourer for the Sunday market. Using money saved from selling empty wine bottles and escargots, they bought a pair of used pants and a shirt.

A few days later, Ali left for Beni Mellal, a small city inhabited mainly by rural transplants. His parents had moved there years earlier because the village had no middle or high school. Ali's older brother, Hsain, arrived on a squeaky old bike a week before school started. Most kids, and even some older men, feared Hsain because of his bloody street fights. He was ten years older than Abdou and had come to take him to register for sixth grade at Moha ou Hammou School.

On Saturday morning, Abdou's mother gave him soap, a threadbare towel, clean pants, and a shirt. "Go to the stream and wash."

The stream was twenty feet from his house, but he went to the winding section in the olive grove for privacy. He stripped naked and bathed in the cold water. While he was washing the soap from his face, Hsain sneaked up, snatched the towel and clothes, and ran.

"I need my pants! I need my clothes!" Abdou yelled, crying like a baby.

Hsain disappeared into the three-hundred-year-old olive trees. Abdou remained naked in the stream until he started to shiver, hiding from passersby. An hour later, Hsain returned, laughing hysterically. He threw Abdou's clothes on the ground.

"Let's get some figs to take to Beni Mellal. Get some baskets."

Abdou was suspicious, nevertheless he went with Hsain.

They climbed the fig trees in their grandparents' backyard, moving from branch to branch like monkeys. They filled the baskets and went home for lunch. Abdou's mother looked at him throughout the meal, tears staining her face. She was going to miss him.

Abdou went to see his goats one last time. He uttered the words softly. "Goodbye, my goats."

After lunch, his mother handed him a small plastic bag with a spare shirt and pants. His father gave him twenty Dirhams—two dollars. They said their goodbyes. His mother began to sob, but his grandmother cut in.

"Stop crying, you fool. You will bring the boy a bad omen. He is going to his uncle's and aunt's house. They will take good care of him."

Abdou fought back tears, trying to mask his crumbling resolve. He was leaving his family at only eleven years old, and it felt less like a departure and more like he was being given up for adoption.

The prospect of the city terrified him. He was heading into the void without his support system—no mother, no father, no brothers, no sisters. He also did not speak Darija-the Moroccan Arabic dialect. Just the unknown waiting for him. He glanced at his father's grim expression and realized the man was struggling to hide his own fear for his son. Then Abdou looked at his younger brother and saw that same terror mirrored in his eyes.

A sharp pang of guilt hit him. Who will protect my brother from the bullies now? he thought.

Abdou grabbed the basket of figs and the bag. He sat on the bicycle frame, clutching his belongings, as Hsain started to pedal. They took a dirt road to the only paved route to Beni Mellal. They hit potholes that felt like they would shatter the bicycle.

The twenty-kilometer trip took four hours, even though it was mostly downhill. They stopped often to eat figs and drink from irrigation channels. They also made two emergency stops behind trees; the boy's intestines were delivering more runs than usual.

They arrived around six p.m. Uncle Addi and Mamma's home was at the edge of the city. The house was unfinished: a kitchen with a fire pit in the center and one single very long room that served as a bedroom, dining room, and living room for fourteen people. Two roofless rooms housed four hens, four goats, and a kid. There was no bathroom, no electricity, and no running water.

Abdou's aunt served coffee with milk, barley bread, and olive oil. As soon as the coffee hit a stomach full of figs, Abdou had the urge to go. He ran outside and unloaded his intestines not too far from the house. It was a near miss.

Ali came out. "That is not where we go. Let me show you."

They jumped a small stream into an orchard. "Your bathroom is behind that tree. The other tree is mine. Each one has their own tree. Just be careful; the owner doesn't like us fertilizing them."

Abdou cried silently. He missed his family. Although the city had more paved roads than his village, he did not like it. It felt less like a city and more like a large, cold village. There weren't enough trees. People did not say hello to each other. They were just a bunch of individuals, not a community.

The next day, Ali and Abdou went to the all-boys Moha Ou Hammou School. It sat atop a hill about four kilometers away. They ran through a winding path in the middle of a vast, thick olive forest.

"Watch for the alcoholics," Ali warned. "If you see them, run for your life. Some are pedophiles. They roam here all the time. Be careful, cousin if they catch you, they may rape you."

The boys registered at school. Abdou paid a ten Dirham insurance fee against damaged property, leaving him nearly broke. His classes included Mathematics, Religious Studies, Egyptian History, Geography, Science, French, Classical Arabic, and Gym. Ali warned him about Mr. Dajaj, the religious studies teacher, and Mr. Mohiddine, the history teacher. Mohiddine was short and wore the same cheap blue polyester suit daily. His shoulders were covered with dandruff. Dajaj had fake teeth. Both punished students who did not memorize the texts.

That evening, Uncle Addi took everyone downtown for supplies. Abdou's remaining dollar wasn't enough for the six notebooks and pens he needed. His uncle paid the rest.

Back home, Mamma served noodles boiled in powdered milk. "Get your clothes ready. Wake-up time is at 6 a.m. You leave by 6:30."

That night, Abdou thought of his parents and siblings. Tears rolled down his cheeks. He missed his mother's food and her "good night." He stayed awake, giggling at the farts coming from different corners of the room, but his mind wandered to the worries of the first day of school.

He fell asleep only to be awakened by a loudspeaker. It was the Imam's call for Morning Prayer, though no one in the house prayed. A few minutes later, Mamma woke everybody. She served breakfast: coffee, toasted bread, and olive oil.

Abdou and Ali ran toward school. The olive forest was eerie in the dawn light. They stopped to relieve themselves, watching for the drunks. The sun began to peek between the trees. It took forty minutes to reach the courtyard. Abdou headed to his sixth-grade classroom. Ali was in seventh grade; he had to repeat sixth grade.

Five minutes before eight, the bell rang. It wasn't a real bell; Mr. Laroussi used a rebar to hit a crop disk, a sound that echoed like a gunshot. The school, once a military base, had classrooms with corrugated roofs arranged around a vast courtyard.

Abdou's first class was math with Mr. Petit, a Frenchman. Abdou was relieved to see Ikhlef, a friend from his village, in the same class. The lesson was straightforward: calculating angles.

Next was French with Madame Demon. She asked each student to describe their summer.

"I watched my goats, hunted doves, and fetched water," Abdou said.

The students laughed. The well-dressed ones boasted of camping on the beaches of Casablanca or El Jadida. Even Ikhlef lied. "I went to the beach with my father and mother."

During recess, Abdou teased him. "Oh, numero 3, you lied. The beach, huh?"

Ikhlef laughed. "The stream near your house and hunting with the slingshot was better than any beach. Anyway, I am glad we are in the same class. With Ali, you, and me, we can face any bullies. Where are you staying?"

"I live with Ali. You?"

"I live with my great uncle."

The bell rang—History and Geography with Mr. Mohiddine.

He started the class with a threat. "Anyone who is late or does not memorize prior lectures will regret being born."

He spoke Classical Arabic with a heavy Amazigh accent. Most people can tell when native Amazigh speakers are speaking Arabic. The class lasted two grueling hours: one on Egyptian mummies, the second on Moroccan rock distribution. Abdou took notes furiously, terrified of corporal punishment. When the noon bell rang, he was relieved. He wished he had never had Mr. Mohiddine as a teacher. He met Ali by the science building, and they sprinted home for a lunch of canned sardines and onions, then ran back for the afternoon session.

    Two weeks into school, Abdou had lost a lot of weight. The running and the poor nutrition were already taking their toll.

CHAPTER 2: Trouble in the House and the Will to Live

Cousin Hsain arrived in the afternoon from boarding school. He was in eighth grade and wasn't expected home until Saturday, even though there was no school on Friday afternoon. His duffel bag, stuffed with all his clothes and books, was a bad sign. It looked as if he had moved out of the dorm completely. His red eyes indicated he had been crying, and the bloody cut on his lower lip suggested he had been beaten.

Mamma jumped up. "Who did this to you? What is wrong?" She ordered everyone else to leave the room, but Ali stayed behind the door to spy.

"What happened?" Abdou whispered to him.

"He was kicked out of school."

There was one more mouth to feed. God help the younger siblings and Hsain's parents. He would torture everyone again.

As the Imam called for the sunset prayer, a man, a woman, and a young boy of Abdou's age arrived. The boy carried a small cloth suitcase. Mamma was unhappy, even though the new guests brought two big bags of meat, vegetables, coffee, sugar, tea, and canned condensed milk.

That night, Abdou and Ali were given a thin gray blanket to sleep on a burlap bag on the floor near the kitchen fire ring. The other fourteen people were crammed into the main room.

"Oh, shit," Ali whispered. "Hassan will stay here as well. Too many people are in the house. Father just lost his job, and now this."

In the morning, Hsain clammed up and sat in the corner as Ali and Abdou left for school. He followed them out and asked them to talk to one of his friends, Rahim, to borrow his notes and homework for Math and Science.

During recess, the boys found Rahim. He volunteered the information immediately. "Hsain screwed up. He was lucky the police did not nab him."

"What happened?" Ali asked.

"He was caught writing on walls after lunch."

"What did he write? We always wrote on walls."

"This is different," Rahim said.

Graffiti was common in the village, usually harmless scrawls.

"Did he write something about girls? What did he really write?"

"He wrote *'Let's go on strike.'* He was caught by one of the assistant principals. He took a beating as well."

That explained why Hsain was lucky not to be arrested.

Last week, there had been too many army trucks everywhere. Something was going on in the country. Soldiers wore helmets and camped near the city center's police station. Walking past the military tents, the boys tried not to look. As children, they were terrified of anything to do with authorities or the state administration. They avoided them at all costs, except when they went to city hall for stamped documents—birth certificates, residence certificates, or the certificate of life.

Morocco ran on certificates; everything had to be stamped and signed. The strangest was the certificate of virginity for women and men—a mysterious document that really just meant a person was not married. However, the boys always wondered how the authorities determined who was granted one.

On Monday morning, Abdou and Ali returned to school with Mamma and Uncle Addi. The parents begged the principal to forgive Hsain and let him return to the dorm. Both parents hardly spoke Arabic; they were Amazigh. Mamma, struggling with the language, tended to address males as females because she had never spoken Arabic before moving to the city.

As they approached the entrance, they saw military trucks. Ali pointed his parents toward the administration building. An assistant principal called the cousins over and slapped each one multiple times—punishment for being five minutes late to class.

When Abdou finally entered the classroom, Mr. Mohiddine, the history and geography teacher, kicked him hard in the tailbone. "Go to the blackboard."

He handed Abdou a piece of chalk. "Draw the location of Egypt."

Abdou drew the map of Africa and the Middle East and located Egypt exactly where it should be.

"Very good," Mohiddine said, and then kicked Abdou again. "Don't be late next time, or it will be worse for you."

Abdou burned with fury and humiliation.

At night, Hassan, the distant cousin, lit a small petroleum lamp his father had bought for him. He placed it directly in front of himself. "This lamp is for me alone."

A skirmish broke out between him and Ali. Mamma came in when she heard the shouting. She told Hassan to share or go to the corner. He went to the far side of the room and lay on the floor to do his homework. Ali and Abdou were furious. The candle they used didn't provide enough light, singeing their eyelashes and hair as they leaned over the small table to solve math problems.

For the next three days, there was no school in celebration of Independence Day. Hassan took the bus to his parents' house in Afourer. Ali and Abdou hitchhiked to the village and walked part of the way.

They enjoyed a tasty rooster tagine with potatoes, onions, tomatoes, olives, and fresh bread, washed down with sweet absinthe tea.

Abdou ran into a few former fifth-grade classmates who had dropped out. He asked them about Rkiya. They all said she could not leave the house now unless accompanied by her parents or older brother. The forty-year-old teacher had requested that she be sequestered to "save her virginity" for him. She was now twelve. She would marry him this summer.

Later that night, Ali whispered, "Hey! I have some black hair growing in my groin. Do you have any yet?"

"No, not yet."

"I heard that if we apply black butterflies to the area, hair will grow faster. Let's find some tomorrow so we can be men."

The two boys wandered the village the next day, searching for butterflies, but found none.

"Where did they go?" Abdou asked.

"They were killed because everyone was spraying their crops with insecticides," Ali said.

After helping Abdou's father clean classrooms, the boys visited their grandparents. On the way back, they grabbed some green mandarins and oranges. The mandarins were tasty, but the oranges were sour; mandarins always ripen faster.

Abdou said goodbye to his parents after a late lunch. His father gave him fifteen Dirhams. They paid twenty-five cents to a taxi that shoved them into the back seat on the laps of two old people. Taxis were licensed for six people—two in the front, four in the back—but this one carried nine, including the driver.

A few kilometers before the city limits, near the Chems Hotel, gendarmes had set up a roadblock with spike strips. They stopped the cab, checked everyone's identification, and ordered the driver out. Abdou peeked through the window and saw the driver handing a blue five-dirham bill to a gendarme. They haggled, and the driver added a Dirham coin to the outstretched hand. The driver got back in, and they continued.

When the boys got home, Hassan wasn't there. They took his little lamp and threw it out of the house. They knew it was terrible, but refusing to share was worse. A few days later, Hassan's parents came and collected him, his books, and his suitcase. He was never seen there again. Unable to bear the living conditions, he moved in with his aunt in Casablanca.

Uncle Addi found a new job constructing a government veterinary service building. He had significant masonry experience, having worked with French companies in Casablanca. Ali was relieved, though the relief was short-lived; the company hadn't paid his father for months. Luckily, it was olive season, and everyone harvested and sold olives for subsistence cash.

Two months later, a barber and a teacher from the village intervened with the principal, and Hsain was allowed to return to school.

Winter break lasted from mid-December through the first week of January. The French teachers went home to celebrate Noel with their families. On the first day of the winter quarter, Madame Demon handed a piece of chocolate to students who could conjugate the verb *manger* in the past participle. Abdou won a piece. He saved it and shared it with Ali during recess.

In mid-January, as soon as the bell rang, a chant began.

"We are on strike! We are on strike!"

Rocks flew at the classrooms. Little by little, most students joined the shouting until the entire school was a massive protest. The principal and assistant principals threatened the students to no avail. Once the military columns started moving in, the kids ran behind the classrooms toward the nearby hill. One kid was caught and taken. A second was beaten and thrown into a giant van.

The students remained on top of the hill, shouting until noon, hurling boulders as military trucks drove down the dirt road below. No shots were fired, and everyone dispersed by the end of the day.

The experience was terrifying for Abdou. It was his first strike. He didn't comprehend what was happening. The next day, he found the school surrounded by the military and the CMI—the Mobile Rapid Intervention Force.

"Be ready to run," Ali warned. "CMI will do anything to preserve order."

Crowds congregated at the entrance. As soon as the bell rang, the shouting began. The CMI moved in to scatter the protesters. Terrified, Abdou and Ali stopped going to school.

Two weeks into the strike, Abdou went to his parents' home. While reading his math book, he overheard his uncle and father discussing the national strikes organized by university students and left-wing parties.

"There were reports of many arrests among university students and union leaders," his uncle said.

This uncle sold vegetables, but he had served in the French army in Indochina, forced to fight a war that wasn't his. After the French left Indochina, he returned and joined the liberation movement. That explained why the villagers found him scary, calling him "Alwatan" (The Nation) or "Resistance."

That night, a respected village elder—also a former resistance fighter—joined the father and uncle for dinner. They ate alone in a room with the door closed, which was highly unusual. Abdou suspected something was wrong. While his mother and grandmother wove a bright red-and-white wool blanket, he went outside near the window to listen.

"There are strikes everywhere."

"They are looking for the grandson of the Haj."

"He ran away from the university. They almost got him. He is brilliant and is supposed to graduate this year with a degree in philosophy. The first from the village."

"What is philosophy?"

"Something they study to learn how to think."

"Haj is asking everyone to hide and protect him. He went into hiding. They may come for him to force the kid to surrender."

"The kid is moving from place to place. Our turn may come to hide him."

"May God protect them."

The next morning, as Abdou prepared to go to fetch water from the springs, his mother handed him a bag containing hard-boiled eggs, bread, sardines, and a stained envelope.

"Give this to a young man who will call you 'Son of Mo'."

He rode his donkey to the springs. A young man in a hoodie and a black djellaba stepped out from the trees. "Son of uncle Mo."

The face was unfamiliar. Abdou handed him the bag.

"How is the family?" the young man asked.

"They are fine, I guess. But they are worried about you and Haj. Something I don't understand."

"Are you still on strike?"

"Yes."

"Continue to resist. Be men and be united."

The young man disappeared among the olive trees.

Abdou returned home. His mother met him in the courtyard. "Did you deliver the goods?"

"Yes, Mama."

"How is he?"

"I could not tell. His face was hidden. He told me to resist and be a man, but I don't know what 'resist' means, Mama."

"Just like your uncle when he was a resistance member."

"Do not talk about what you saw today. It will be dangerous for him and for us. Understand?"

"Yes, Mama. All I want is to return to school. I don't want to be like my uncle. I like math. I am tired of being poor and of what I see every day." Tears filled his eyes. He was afraid of the knowns and the unknowns.

She smiled, put her hand on his head, and pulled his curls. "You are returning to the city after lunch. Your father received a severe letter from the authorities instructing him to bring you back. If things do not change, your time will come."

He did not know who the young man was, or what she meant by "your time will come."

After lunch, Abdou and his father took a taxi to Moha ou Hammou school. The military and police presence was overwhelming; trucks were parked inside the school grounds. A soldier directed them to a large tent where a line of parents and students waited.

When they reached the table, a tall man with scary eyes and a thick black mustache shoved a paper forward. "Put your finger on the ink and print here. Mo, you will be jailed if your son strikes again."

His father signed without reading.

Abdou kissed his dad's hand and went to his classroom. He found Ikhlef and a few others, but many seats were empty.

"The strike is over, Abdou. Welcome back," said Mr. Petit, the math teacher. He handed Abdou a few sheets of missed lessons. "Finish the problems this weekend. Quiz on Monday."

When the bell rang, Abdou hurried to find Ali, but his cousin's classroom was empty. It was getting dark and cold. Abdou hurried out, avoiding eye contact with the police. He found some of Ali's friends and ran with them through the dark olive forest, heart pounding.

When he got home around seven, Mamma was surprised.

"Did you just get here from the village?"

"No. Father took me to school. He had to sign. He will go to jail if I strike again. Did Uncle Addi take Ali to sign?"

"No, your uncle got a job building apartments for the government in Khouribga. I will take Ali tomorrow."

Ali, Mamma, and Abdou went to school early the next morning. Mamma was terrified. She was fingerprinted and told the same threat:

jail if her son struck again.

A week later, slogans started as soon as the bell rang. But this time, the police and the army intervened immediately, dispersing the students with force.

At night, Abdou lay on the floor reading a poem in his Classical Arabic textbook. It was called "The Will to Live" by the Tunisian poet Abou Al-Qacem Echebbi.

If, one day, a people will to live,
Then fate must answer their call;
And the night must fade away,
And the chains must break and fall.
He who is not embraced by longing for life
Will dissipate in its atmosphere and fade
Woe to him whom life does not shake—
He shall vanish into eternal shade.

Abdou smiled. He realized this was one of the slogans his classmates had been chanting. He finally understood what the young philosophy student meant by "resist," and why people called his uncle "Alwatan."

His uncle often talked about the will of the people and breaking chains. Now, Abdou knew what those chains looked like.

CHAPTER 3: A Busy and Scary Summer

School ended in June 1971. Mr. Petit congratulated Abdou for doing very well in his sixth-grade class and handed him a seventh-grade math book as a prize.

"Congratulations, little Abdou. Read this book during the summer and practice all the problems at the end of each chapter," Mr. Petit said.

Madame Demon gave him a thick book titled *Histoire de la Révolution Française*—The History of the French Revolution. Mr. Mohiddine, the jerk, gave him nothing, even though Abdou had scored 19/20 in his class. Mr. Dajaj, the religion teacher, told him he had barely passed. Abdou was embarrassed by his near failure in religious studies, but he consoled himself with the fact that religion was only two credits, while math was five.

"You are done with school. Now you have to go to the village and water the orchards," Uncle Addi and Auntie Mamma said.

Abdou and Ali went to the village. The two were inseparable, mainly because neither had yet made good friends in the city.

People in the village took turns watering their fields. Not everyone had water rights, and certainly not equal rights. In the village, water rights were inherited just as land and property were. Some families owned the entire stream flow for four hours; others had rights to only one-quarter of the flow for an hour. Elders strictly monitored and enforced these unequal rules. Tension was constant. Thieves would sometimes divert water at night to irrigate drying fields, leading to bloody fights among neighbors.

A week after school ended, while waiting for dinner, Abdou's father pulled an open letter from his pocket.

"Congratulations, son. You did very well, except in religious studies." He paused. "You will start studying at the mosque tomorrow."

"But I don't want to," Abdou said.

"You do not have a choice. You need to improve your knowledge of religion. You almost failed it."

Ali and Abdou began going to the mosque to study the Quran. They started at seven a.m., went home for breakfast at nine, and left the mosque at eleven. Students took turns bringing breakfast and lunch to the Imam. On Mondays—market day—they also brought him an egg or five cents as pay. Sometimes, the Imam left the kids alone while he went to have tea at neighboring houses.

The Imam gave each student a wooden slate coated with wet marl, a soft, white-gray sedimentary rock made of calcite carbonate and clay. The students wrote verses of the Quran on the slate using a sharp bamboo stick dipped in ink made from burned lamb wool and water. The ink washed off easily. The students, known as *Imahdarn*, had to memorize and recite the verses with their eyes closed before erasing the slate to start the next set. Anyone who failed or changed a single word of Allah was whipped with an olive stick. The children were sick of olive branches.

The Imam enforced discipline by inflicting pain on the pupils.

A few weeks into the summer, the Imam returned from one of his tea visits to find a few older kids playing marbles inside the prayer room. He was furious, and everyone was punished. Rumors swirled that the Imam spent too much time at a young widow's home after midday prayers, committing a sin that God did not like.

That evening, the elders met and decided to expel every student aged 12 or older from the Quranic School. They decided the older boys were a bad influence on the younger children.

"Ali, this is great! We don't have to endure the Imam's wrath," Abdou said.

"Yep. We will have plenty of time to hunt. We should head to Takarrout Mountain tomorrow. The jujube trees have a lot of doves and Spanish sparrows."

Their conversation was cut short.

"Now you have to do more chores," Abdou's father said. "I wanted you to learn the word of God, but you messed up. Tomorrow, go with your cousin to gather straw for the cows, lambs, and goats. We need animal feed for the winter."

In the morning, Abdou's father shaved Abdou and Ali's heads to prevent straw and lice from sticking to their hair. His hand was not steady, and the dull shaving knife gouged chunks of skin from their scalps, leaving permanent scars.

"Your head looks like a wounded testicle," Ali said.

"Yours too, cousin.'

Every day, the boys rose early to gather straw, filling the storage room in front of the house. By 10 a.m., temperatures soared to 120 degrees. The boys sweated buckets. After each load, they ran to the stream to wash and cool off. In the afternoon, they napped to avoid heat stroke.

Later, they roamed the village, hunting with slingshots fashioned from Y-shaped olive branches, old shoes, and inner-tube rubber strips. Sometimes, they went to the city and rummaged behind the hospital to find discarded surgical tubing; it worked better and fired a shot like a bullet.

Once the storage room was full of straw, Abdou hatched a plan. He visited old ladies and bought eggs for three cents, reselling them for five or six cents on the dilapidated paved road that crossed the village.

The margins were good, but the demand was too low to justify the time. They decided to sell only on Mondays, the day of the Souk.

They sold their eggs early in the morning for five cents each to a middleman—a bald poultry reseller with gold teeth and a sparse mustache who tried to fleece them at every turn.

"Asshole. He thinks we know nothing," Ali said.

"Yeah, I hate his gold teeth. The city man comes here with his pickup truck to the Amazigh country, thinking, *Ah, these hicks know nothing*. But let's start selling him chickens. We need the money."

"We don't have enough chickens at home, and we don't have money to buy more chicken stock," Ali said.

"We can ask the old widows if they want us to sell their chickens in the market for them. We find the ones who cannot walk far. Everyone wins. We'll start with the neighbors."

They went from house to house. Two widows gave them six roosters and four hens.

"I need at least twenty Dirhams for all my chickens," one woman said.

"We'll do our best to get you a good price," Ali promised.

They sold the birds to the old gizzard with the gold teeth. In a few hours, they pocketed fifteen Dirhams commission. They paid the half-Dirham tax to the soldier enforcing market rules and gave the asking price to the widows. By the end of the summer, Ali and Abdou had amassed one hundred fifty Dirhams—more than Abdou's father's monthly salary.

One day, they took four roosters to the market. The taxman and his soldier showed up before the man with the gold teeth arrived. The boys hadn't sold a single chicken yet and didn't have a half Dirham to pay the tax.

The soldier reached out to confiscate a rooster. Abdou grabbed the bird by its feet. The chicken flapped its wings violently, slapping the soldier in the face. Furious, the soldier grabbed Abdou by the collar and dragged him to a windowless room where the *Cheik*—the village's highest authority—rendered verdicts.

Ali ran to find Uncle Haddou, who was selling vegetables in a tent. Abdou, detained and terrified, was about to pee his pants. The soldier slapped him and shoved him to the back of the room.

Uncle Haddou burst in, yelling at the Cheik.

"Calm down, Mr. Haddou. Calm down, *Resistance*," the Cheik said, using the uncle's nickname. "What is wrong?"

"How can I calm down? You arrested my nephew! The son of Mr. Mo and my sister Fadma." Haddou turned to the soldier. "And you! I wore that uniform to liberate this country before you were born. Have you no respect for my family?"

"I am sorry, Resistance! Forgive me," the soldier stammered. "I did not know he was the son of Mo. I respect his parents; they are poor folks like my family. I did not recognize the boy."

"You can leave, son. This is a misunderstanding," the Cheik told Abdou.

Abdou grabbed his chickens and ran, selling them tax-free. It was his first arrest and his last act of tax evasion.

"Mom, Uncle Haddou is brave!" Abdou said later. "He is not scared of authorities, unlike Ali and me."

Uncle Haddou stopped by later that day. He yelled at Abdou, "Don't get in trouble again! What if I hadn't been there? Avoid the authorities. Luckily, the Cheik's brother was my brother-in-arms. Together, we burned down a French outpost in 1950. But my dear friend, the Cheik's son, was killed when a traitor reported him. I burned down the traitor's house and fields. That is why the Cheik respects me."

On July 10th, 1971, the summer took a dark turn. Abdou and Ali were listening to popular Amazigh music on the radio when the broadcast cut out. An enthusiastic, terrified voice began to shout.

"The army is revolting! The revolution has started!"

The boys ran to the roof, turned the volume up, and shouted, "The army is doing a revolution!"

An older neighbor scrambled up the ladder and grabbed the radio, snapping it off. He shook the boys hard. "Do you want them to come and kill us all? Go downstairs!"

He checked to ensure no one else had heard. He dragged them inside to their mother. "They were shouting that the army started a *coup d'etat*."

"What is a *coup d'etat*?" Abdou's mother asked.

"The army is trying to remove the king."

"Why?"

"Corruption. Poverty. That was why the kids were striking, and why Haj's son is on the run. Let's pray God spares our country a bloodbath. Our enemies in the East and Gaddafi in Libya would be happy if we started killing each other."

The neighbor, a veteran who spoke fluent German from fighting Nazis in World War II, looked grim.

The family stayed glued to the radio until dawn, surfing between stations. BBC Arabic, Radio France International, and Radio Tangier gave conflicting reports. The Algerian and Libyan stations cheered for the King's death. Abdou and Ali struggled to translate the news.

The night before, a thousand military cadets from the training school in Ahermoumou had attacked the King's birthday celebration. Many were killed. King Hassan II escaped, and his loyal General Ben Bouhali was killed putting down the mutiny led by Colonel Ababou.

It was a terrifying time. Months later, the mutineers would be executed.

Abdou finished the math book Mr. Petit gave him. With Ali's help, he solved every problem. But he only finished half of the French Revolution book. It was dense, and he didn't understand why the French were revolting, though the illustrations of half-naked women were distracting.

He managed to grasp that the French revolted because of poverty and abuse by their King. He thought of the violations by the Imam, the slap from the soldier, and the terrifying voice on the radio. Abdou kept thinking about the French motto.

Liberty, Equality, Fraternity.

Did they really mean it? Was that slogan only for the French, or for the people they occupied? Was it for immigrants? Or was it just words, like the Imam's prayers? The same Imam who was fired by the community to bring an Imam who followed the faith.

CHAPTER 4: The Eviction and the Wedding

On August 27th, 1971, Abdou helped his mom catch a big rooster for lunch. He took the bird to a neighbor to be butchered. The neighbor pinned the rooster's wings under his right foot, muttered a prayer, and sacrificed it. Abdou looked away from the gushing blood. He carried the limp bird inside, dunked it in a pot of scalding water, plucked the feathers, cleaned it, and handed it to his mother, who was already peeling a mountain of vegetables.

He then went with Ali to his grandparents' house to fill a basket with tomatoes, zucchini, red squash, hot peppers, and potatoes fresh from the farm to take to the city.

When they returned, they feasted on delicious couscous with chicken and seven vegetables, the customary Friday meal. In Morocco, you didn't mess around with couscous; it wasn't real unless it had at least seven vegetables. It was a sacred tradition.

The boys left for the city on an old, rusty bike, dropping off the vegetables at Ali's parents' house before heading downtown to buy clothes from Khalla, a shop owner from their village. They stuck to Khalla's shop because their Moroccan Darija was limited, and Khalla spoke their Tamazight language.

To get there, the cousins had to walk through Hnajra, the narrow prostitution street not far from Khalla's shop. Old and young men haggled openly with women for services. Abdou and Ali burned with embarrassment. They kept their eyes on the ground as they passed women exhibiting their bodies in short, revealing clothing. The walk through the winding street felt like an eternity, even if at times they enjoyed the view and had many questions.

The teen boys bought pants, shirts, and cheap tennis shoes. They didn't buy socks or underwear; they only wore underwear when they went to the public bath, the *hammam*.

"Ali, tell your dad that he needs to vacate the house," Khalla said as they paid. He was not just a shop owner but also a rental agent for a sleazy absentee landlord who dabbled in rundown properties. "The owner and the neighbors complained about you. You have too many people in the house, so you are being evicted."

"Yes, Mr. Khalla. I will tell my mom because my dad is working in Khouribga. He won't be back until next month or Ramadan," Ali said.

"The owner and I don't care. You must leave the house within three days, or I will bring the *Makhzen*."

The threat of the authorities hung in the air.

They left the shop with heavy hearts to deliver the bad news to Aunt Mamma.

"Khalla, the son of a… took advantage of us and threatened us," Ali muttered to Abdou.

"For sure. What will happen to us now that we are being evicted?"

"God knows. I don't want us to leave school and go back to the village. It is enough that your father made us gather straw all summer."

"You idiot. You and Ikhlef played marbles at the mosque, so we got kicked out and punished," Abdou reminded him.

"We will leave the matter to the Almighty. If he listens to poor people like us. I feel sorry for my mother."

When Mamma received the terrible news, she didn't panic. Though she spoke no Arabic—only Tamazight—she was hardened by village poverty. She acted quickly.

"Find us a place," she ordered the boys.

The eleven boys fanned out across the city's neighborhoods at the edge of the town. They asked strangers, Amazigh shop owners, and schoolmates if they knew of a rental. It was a massive, collective search. Within a few hours, they found a place.

It was on a busy street, owned by a skinny, short, old widow.

"You have an accent; where are you from?" the lady asked.

"We are from Timoulilt, but we live in Oulad Ayad," Ali said.

"You are Amazigh? I know people from Timoulilt. Why are you moving? Where is your dad?"

Abdou jumped in quickly. "Our father got a new, better job. He works out of town."

Ali added a lie for good measure. "My father is the chief of a construction project for OCP."

It was a bold claim. OCP was the massive state-owned mining company that was the world's leading exporter of phosphates. In reality, Uncle Addi was no chief, just an excellent mason.

The skinny widow was convinced. "How many people are in your family?"

"Parents and another couple, brothers and a sister. This is my cousin," Ali said.

"The cost is sixty Dirhams per month. Bring the cash tomorrow. Two months' rent in advance. You will not get any receipts from me, but you have my word as long as you pay on the first of the month."

Mamma and Hsain went to the gold market downtown, where she sold two pieces of jewelry. Abdou and Ali went to the farmer's market and sold the goats for 55 Dirhams each. In those days, husbands bought jewelry or livestock as an emergency savings account. Mamma sold the gold for 130 Dirhams. Hsain took the cash and paid the landlord, securing the keys.

On their way back from the market, two older thieves ambushed

Abdou and Ali. Abdou pulled a bicycle chain from his pocket and whipped one of the thieves hard. Ali drew his gravity knife from his back pocket and swung it at the other. Both robbers ran away screaming. The cousins didn't wait around; they sprinted all the way home. Beni Mellal's markets were full of thieves, but Ali and Abdou had been training in street fighting since they were nine. Ali had taught Abdou the golden rule: kick first, hard, and repeat fast.

The new house was a kilometer away. It had two rooms, a kitchen, a courtyard, a bathroom, and running water, but no electricity or stove. It did, however, have a spacious flat roof.

After lunch, Mamma announced the plan: they would move at night to avoid drawing attention in the new neighborhood. They borrowed a pushcart with bicycle tires from a neighbor.

Under the cover of darkness, the migration began. The boys walked in single file, taking shortcuts through olive and orange orchards. They carried pots, pans, two large serving plates, the tagine, a large couscousier, spices, oil drums, and flour. Ali and four of his brothers made extra trips for the blankets, pillows, and mats. Mamma and her daughter set up the new house immediately and started dinner. Abdou went to buy candles and petroleum oil for the lamps.

Everything was moved in a few trips. Abdou and Ali struggled with a heavy table Uncle Addi had built from discarded construction wood, covered in white tiles stolen from the veterinary service site.

By midnight, Mamma served a delicious stew made with vegetables from the village. The family ate in exhaustion and went to bed by 1 a.m.

In the morning, Mamma sent Hsain to buy a bag of coal since the house lacked a fire pit. Abdou and Ali went to Mai's junkyard and bought a small oil drum. The owner cut it down to make a brazier and welded two pieces of rebar on top for pots and kettles. Mamma was thrilled. She set it up in a corner of the roof and used it to bake pita bread on a thick clay plate.

One cousin went to buy eggs—his younger brother had broken the entire basket during the move. The other children took a hatchet across the street to the large olive field, gathering enough dry wood to last weeks.

After a lunch of eggs, tomatoes, and onions, four of the cousins hauled bags of dirt from a clay hole. They mixed the clay with wheat straw and water to build an oven on the roof. The wet clay kept collapsing until the brothers built a mound of rocks, covered it with straw and chicken wire, and plastered over it. They left a small opening for smoke and a large one for wood and dough. In the grueling August heat, it dried in a week. Once Mamma removed the rocks and lined the bottom with broken clay pots, she baked the most flavorful bread they had ever tasted.

The family settled in. The new house was comfortable, though they missed the milk from the goats they had sold. The boys understood the new rules: the indoor bathroom was for the women. The boys were to fertilize the trees across the street. Just like in the old neighborhood, they each picked a favorite tree and used pages from old notebooks as toilet paper.

Ali and Abdou returned the cart to the old house and left a note with their new address for the neighbors. Then, they hitchhiked back to the village to attend the forty-year-old teacher's wedding.

"He is going to make Rkiya a woman in a few days," Ali stated.

Abdou became upset. His crush on Rkiya came to the surface. "I wish I could run away with her."

"Then what?" Ali asked.

"Go live in one of the orchards."

"Keep dreaming, little one," said Ali. "You don't even have hair in your groin, you little snot. Get over Rkiya. Fantasizing about another man's wife is *haram*. That is a sin. You will go to hell."

"But she is only twelve," Abdou almost cried.

At night, the toothless teacher came to Abdou's house. He didn't knock politely; he stood at the door and barked orders at Abdou's father.

"Bring your chef's hat and nice clothes because a lot of dignitaries are coming," the teacher ordered in an authoritative voice. "You will cook chicken with roasted almonds. The women will cook couscous."

Abdou watched his father nod respectfully. Some of the teachers like Bami, Darmouch, and Chakir were very nice to him, treating him like a colleague. Others, like the toothless teacher were mean. They considered him a servant, not a government worker, simply because he was a janitor and an outsider to the village. The "foreigner" label stuck to anyone not born in the locality, creating a caste system that Abdou despised.

The wedding was a spectacle. Many other teachers, local authorities, and big landowners attended. A music troupe entertained the guests, the drums echoing through the valley.

The party was strictly segregated. Men sat in a large tent outside the house, while the women gathered in another tent nearby. They did not mix—except for the professional female dancers (the *chikhates*) who performed for the men.

One young landowner had too much wine to drink. He danced wildly until he collapsed on the ground, unable to stand. He tried to crawl, but his legs failed him. Two guests finally picked him up and carried him to a corner, where he passed out and snored loudly right next to the Imam for the rest of the night.

The uninvited guests, including many of Abdou's friends, gathered outside the tent. They watched the dancers and singers, who performed with a freedom that made the adolescents' groins swell, and their imaginations run wild, dreaming of lives far beyond what they were used to.

"I like that dancer," Ikhlef said. "She bounces like a young mare. Look at her small breasts bouncing like oranges."

"No, I like the older one," Ali said. "I like her brown thighs. I could caress them all night."

"Look at the teacher," Ikhlef whispered. "He is moving like an animal."

Abdou realized the teacher was celebrating the end of a child's education and life. The man came out of the house all smiles, proud of himself.

Soon after, the women's tent filled with yodeling and songs. The adolescent group ran toward the tent to see Rkiya's aunt carrying a tea platter on her head and shouting triumphantly: "Daha oh! Daha! Oh! Allah Ma Khallaha" – meaning he took Rkiya and he did not leave her.

The boys moved closer to peek from behind the tent. Rkiya's older brother came running at them. He threatened to kick Abdou but stopped when he saw Ali. Instead, he launched a series of insults.

"You son of an outsider!"

"You son of the soup man and janitor!"

"Go to your home. All of you go to your homes. No one invited you." He turned to the others. "You too, Ikhlef. Get the hell out of here. You son of landless peasants, get out."

The women continued to yell and sing until past midnight. The guests congratulated the forty-year-old teacher, saying, "*Mbarak Masood*"—congratulations.

Abdou was hurt and angry at what the forty-year-old man had done. He finally accepted that Rkiya was gone forever. He went home very late.

The next morning, Abdou was upset, but he distracted himself by hunting. He bagged six doves. He cleaned them and made a dove stew. His grandmother warned him that doves would give him excessive farts, but he did not mind. At least he got some meat that day.

But the thoughts of Rkiya and the old teacher never left him. Questions swirled in his mind, leaving him unsettled.

The teacher put an end to Rkiya's education.

Will Rkiya have a baby?

Why did Rkiya's family let her down?

Why did no one save her?
Did all the dignitaries and the Imam approve of this?
Where was the law? Wasn't it supposed to protect children?
The summer break ended quickly. The forty-year-old teacher never paid Abdou's father for the work he did during his wedding.

CHAPTER 5: The 1972 School Year

Ali and Abdou helped clean the elementary school in Timoulilt to get it ready for the new term. Then, they packed their clothes and returned to the city to register for the 1971-1972 academic year. This time, Abdou had saved enough money from his egg-and-chicken reseller business and from selling snails and aromatic plants he collected.

However, due to the destruction caused by students at Moha Ou Hammou High school during the strikes, the administration had raised the registration fees from 10 to 35 Dirhams. After paying the fee and buying school supplies, Abdou had only about ten Dirhams left. The boys realized they would have to search for additional income during the school year.

Abdou started 7th grade on September 20th, 1971. The routine was familiar: Mamma woke everyone up at 6 a.m. The kids washed their hands and faces, ate bread dipped in olive oil, gulped a small glass of black coffee, and ran out the door to climb the hill, aiming to reach school by 7:30.

Hardly anyone attended class on the first two days. Most families couldn't afford the fee hike, so the administration postponed the start date by a week to give parents time to find the money. Abdou and Ali took advantage of the delay to buy and resell cactus fruit. They went to the farmers' market early, bought a case of prickly pears, and set up shop near a creek in the city center.

To their chagrin, before they could finish selling the case, the *Makhzen*—security forces charged with harassing street vendors—rushed in. They kicked the crate, dumping the remaining fruit into the muddy creek. They tried to grab the cash box, but Ali was too fast; he snatched it and sprinted away.

A week later, classes finally began.

Abdou had a roster of new teachers: Mr. Assadi for classical Arabic, Mr. Bailleul for mathematics, Mr. Gloriel for French, and Madame Petit for biology and science. Madame Petit was married to Abdou's 6th-grade math teacher. She was a tiny French Catholic woman in her late thirties with brown hair and a pale face—features that looked soft compared to the hardened, beautiful faces of the village women.

The first lesson was about the life cycle of chickens.

"What do you know about chickens?" Madame Petit asked.

A few kids shouted out answers. "Delicious, Madame!" "Scary!" "Noisy!" "Stinky!"

Abdou thought, *Moneymakers,* but he remained quiet. With his experience in the village, he knew everything about chickens. He knew that a rooster could mate with three or four hens in a single minute.

Madame Petit turned toward him. "Abdou, what do you know about chickens?"

"A lot, Madame."

"Like what?"

"You see, Madame, the clothes I am wearing and all my school supplies... I bought them because I sold chickens and eggs this summer. I even got arrested because of chickens. I also raise my own."

She laughed a bit. The entire class started to laugh, pointing at him. "Chicken Boy," someone whispered.

Madame became impatient. "Silence, *s'il vous plaît!* Abdou, tell the class about the life cycle of a chicken."

"*Oui*, Madame. A rooster mounts the female chicken, giving it its seeds in its cloaca."

The class giggled at the word *cloaca*, but Abdou continued, smirking.

"The female chicken lays eggs. Once it lays enough—usually fifteen to twenty—it sits on them for twenty-one days to keep them warm with its body temperature. After that, the little chicks break the shells and come out. Not all chicks hatch. The mother becomes very protective, but some chicks die a natural death. Others get eaten by hawks, weasels, or raccoons. Most survive, grow, and become adults to lay eggs or be eaten by humans."

Madame Petit clapped. "Bravo, Abdou. Marvelous. In the next class, we will cover the biology of the egg and the reproductive system."

Math class was a breeze since Abdou had read the entire 7th-grade book over the summer. The classical Arabic teacher was easygoing; he assigned a text and took naps while the kids read, answering questions only after he woke up.

Mr. Gloriel, the French teacher, was a different story. He was intense, often rolling up his sleeves to show off his biceps and intimidate the children. He always smelled of alcohol, his eyes red and glassy. His lips were often stained with his wife's lipstick. The French usually had wine with lunch, but Mr. Gloriel seemed to have had the whole bottle.

He treated his pupils like little savages, acting as if France still occupied Morocco. Anyone who made a spelling mistake was treated to a slap in the face or a kick in the stomach. Abdou avoided trouble, but he was terrified every time he entered the classroom.

Mr. Gloriel never drove himself to school. His wife dropped him off. Both in their thirties, they would get out of the car—Madame Gloriel often in a mini-skirt that exposed her legs and chest—and hug and kiss passionately before she drove away to another high school.

Hordes of kids waited for the Gloriels every morning just to watch the French kissing and Madame Gloriel's figure. The couple seemed aware of their audience and would smooch even longer, driving the hormone-filled teenagers crazy. Assistant principals sometimes stepped in to disperse the crowd, but often, the educators stood back and watched the show too.

Ramadan started on October 20th, 1971. During the holy month, Muslims fast from sunrise to sunset. Mr. and Mrs. Gloriel had to stop their public lip-sucking out of respect—or perhaps fear—of the conservative atmosphere, much to the students' disappointment.

The city streets became deserted between 11 a.m. and 6 p.m., only to explode with life and noise after the sunset meal of dates, figs, and *Harira* soup.

During this time, Uncle Addi returned from his extended stay in Khouribga. He was surprised to find the old house locked and empty. A neighbor guided him to the new rental. He brought a bounty with him: vegetables, meat, *chabbakia* pastries, and dried fruits. He gave Mamma money, but he looked exhausted. Working with cement had taken a toll on his hands; his fingers were cracked and raw. Employers in Morocco never supplied gloves.

Addi reminded the boys to study hard. "Become engineers for OCP," he told them. "They have free housing, big green yards, and cars."

The *Eid* celebration coincided with Independence Day on November 18th. Abdou returned to Timoulilt to celebrate. The village road was decorated with palm branches and banners reading "God, the Country, and the King."

To Abdou's surprise, his father introduced him to a stranger. "Meet your nephew, Mohand."

Mohand had just appeared in the village. He had lost his parents to tuberculosis in a remote town. He looked healthy but different—he had very white skin, smooth hair, and looked more European than Moroccan. He was already shaving. Abdou said hello but kept his distance.

"You know that you cannot walk inside the olive orchard," Abdou's mother warned him. "The ban is enforced by the *Izmazen*."

The *Izmazen* were the twelve village elders designated to protect the harvest. No one was allowed to enter the orchard or bring herds in until the *Izmazen* gave the word. Violators were punished by having to feed dinner to all twelve elders—a costly fine.

A day after Eid, the *Izmazen* fanned out, announcing the harvest. The next morning, the orchard came alive. Men climbed the 300-year-old trees, praising God as they beat the branches with long bamboo sticks. Women and children collected the falling fruit. The olive forest was full of sounds "Allah Akbar! Salla ala Sidna Mohamed" – God is the greatest, prayers on prophet Mohamed.

The Olive Festival lasted two months. Vendors lined the roads, buying from farmers and selling to rich intermediaries who exported the high-quality oil to Spain.

Nephew Mohand started work immediately, but his luck ran out quickly. He fell from a tall tree and broke his arm. Grandma took him on a donkey to a healer in a neighboring village. Two men held Mohand down while he screamed as the healer set the bone. He returned with a cast made of split bamboo and lamb's wool, tied together with a goat-hair rope, his arm supported by Grandma's shawl.

School resumed, but everyone was waiting for the winter break. Abdou spent his free time hunting and selling olives. He encountered a vendor known for cheating people. The man used an old metal scale with a magnet attached to the bottom, stealing about 250 grams per kilo.

Abdou sold the crook some olives, then bought a can of sardines from the man's cart. The crook opened the can, quickly stole a sardine, and handed it over. Luckily, Abdou hadn't paid yet. He glared at the man, watched him steal twenty-five percent of his lunch, and simply walked away.

Scarcity pushed people to the edge. The poor cheated the poor; the rich cheated everyone.

By the 1970s, the United Nations was working with the Ministry of Health on a family planning program. Women with more than three kids were subjected to IUD implants. But the staff was poorly trained, and the multi-thread devices used were often defective, dumped on African and Asian nations by companies avoiding lawsuits in America.

In the village, rumors floated that the devices were harmful, or even against Islam. Darker rumors swirled about Rkiya. People whispered that she had bribed someone to implant an IUD in her because she refused to bear children for the forty-year-old teacher who had taken her childhood.

The final exams ended on December 14th. On the last day, Madame Petit brought cookies to share. She handed Abdou a thick book titled *The Biology of All Living Things*.

"Read a few pages every day," she told him. "You did marvelously this trimester."

Abdou took the book home, his parents beaming with pride.

CHAPTER 6: Crime and the Angel

Uncle Addi lost his job again because he had stayed in the village for nearly two months to harvest olives, plant beans and peas, and sell the orange crop. The financial situation was dire. Mamma managed whatever money was left, but everyone in the house understood they were destitute. They ate whatever was cooked without complaint, even though the portions were always insufficient.

Addi eventually found a new job building a hotel in Beni Mellal, but the contractor failed to pay him for almost four months. Desperate, Addi made a deal with a coworker who was moving south to care for his aging parents. Addi paid him 300 Dirhams for his "squatting rights"—two rooms on a farm abandoned by French settlers after independence.

The family moved in. There was no electricity, running water, or a kitchen. The "house" consisted of two rooms surrounded by an acre of brush and barbed wire. The place was infested with ticks and rats, but at least there was no rent to pay—until a connected landowner eventually falsified papers to kick them out.

Uncle Addi built a makeshift kitchen from two-by-fours, plastic sheets, and corrugated metal. He also built a tiny bathroom for the women—a bucket behind a screen, emptied a few yards away.

Cousin Hsain came home from boarding school on Saturdays. He treated the family's new plot like his personal fiefdom. He forced everyone to extend the garden every weekend. He took Abdou, Ali, and cousin Brahim to the edge of the olive orchard to cut and haul brush with long, deadly thorns for fencing.

While the younger boys worked, Hsain sat under a tree, reading his biology book. He acted like a slave driver. Each kid had to fasten a pile of brush with a rope and drag it for kilometers. Anyone who stopped for a break got a whipping.

One weekend was nearly fatal for Abdou. He and Ali were carrying a heavy olive tree trunk on their shoulders. Ali stumbled and dropped his end; the trunk bounced and slammed into Abdou's head. He fell unconscious for several minutes. When he woke up, dazed and hurt, Hsain forced him to finish carrying the trunk home.

When they finally arrived, Mamma asked Abdou to fetch water from the public faucet a kilometer away. By the time he returned, everyone had eaten. There was no food left.

The misery continued until Abdou decided to stop coming home on Saturdays. Instead, he left school directly for his village to escape Hsain's forced labor camp. But Hsain caught him. He picked Abdou up and threw him face-down into the thorny brush.

"You keep leaving directly from school because you are avoiding work!" Hsain yelled.

"No, I went to see my parents and Mohand!" Abdou cried.

He showed up to class on Monday with his face scratched and bleeding.

"What happened to your face?" Madame Petit asked as he sat down.

"Nothing, Madame."

"Who did this to you?"

"I fell, Madame."

"Did someone push you?"

"No, Madame. I was trying to get an orange from a tree and fell." Abdou lied, terrified that Hsain would punish him if he told the truth.

Madame Petit didn't believe him. Later, she came by with her husband and applied ointment to his cuts.

"How are you, little Abdou?" Mr. Petit asked.

"*Ça va*, Monsieur Petit."

"We will be driving through your village on Saturday to go fishing in the lake. We can drop you off at your parents' house and bring you back on Sunday."

"Thank you, sir, but I have a lot of homework, and I promised my cousins I would work on the yard. If I go, my older cousin will beat me again."

Madame Petit froze. "So your older cousin is the one who beats you?"

"Oui, Madame. But please, I will get in trouble if he finds out I told you."

"Is he the one who scratched your face?"

Abdou did not respond. He stared at the ground like a dead man.

Fearing another beating, Abdou stayed that weekend. He spent two days cutting brush and dragging it for kilometers.

Once the garden was finally fenced and planted, Hsain found a new use for the boys. He started taking Abdou and Ali to the city center, forcing them to steal from farmers bringing goods to market.

The kids shadowed farmers riding donkeys or bicycles, snatching whatever they could—clothes, vegetables, meat. Hsain took the loot home and told Mamma that her brother had sent the goods as a gift.

Emboldened, Hsain took the kids to a crowded market known for pickpockets. His orders were to steal wallets and money. Hsain stayed at a safe distance, watching like a hawk, collecting the stolen goods after each theft.

Ali was caught once. The crowd descended on him instantly. In Morocco, when someone shouts "The thief!", everyone joins in to beat the culprit. Ali barely escaped with his bones intact before the police arrived.

Hsain decided to change strategy.

"We will no longer steal directly from people," Hsain announced. "We will rob the thieves."

It was a twisted pyramid scheme. Ali, Abdou, and Hsain would identify pickpockets, follow them, and confront them after a score.

"You thief! You stole from my father!" Hsain would yell, grabbing the thief's loot.

It worked until they targeted a hardened criminal who pulled a knife. Abdou, terrified, refused to go on any more missions. Hsain beat him for his cowardice.

Life became unbearable. When spring break ended, Abdou refused to return to school. He wanted to quit and work on the farms.

His father sat him down. "Son, you know we are destitute. I want you to finish your studies. Life is tough without an education. Look at me. Look at the people around us. Everyone is barely getting by."

"But Dad, life is tough at Uncle Addi's house. There isn't enough food. I cannot study on the weekends. Hsain treats us worse than slaves. He forces us to steal. Look at the scars on my face."

"Son, you don't want to be someone else's slave forever. I have big hopes for you. Ask God for forgiveness for the stealing. Do not steal anymore."

Abdou knew what slavery looked like. He had seen the sharecroppers—the *Ashrik*—in the village. They weren't in chains, but they were owned in every other way. They lived with the landlords their entire lives, forbidden to marry, eating scraps in the courtyard like dogs, working day and night.

"Refuse Hsain's orders," his father said. "If he beats you, come back home. I will buy more sheep and goats for you to raise. But try one more time."

Abdou agreed.

On Monday, Madame Petit handed Abdou a folder after class.

"I know you have a hard life, little one," she said softly. "This is an application for a scholarship to the boarding school. I filled out most of it for you. You have good grades, but I want you to break a record. My husband and I wrote letters to the principal, Mr. Alami. He promised that if you beat the record, he will defend your application."

"Merci, Madame. I will try my best."

Abdou felt a fire light inside him. He was determined to escape the hell of Hsain's rule.

Fate intervened in his favor. Hsain's grades had dropped so low that the boarding school detained him on weekends for mandatory study hall. The school sent a letter warning his parents that he risked losing his scholarship. It was bad news for Mamma and Addi, but salvation for the kids at home.

Abdou studied relentlessly. He stopped playing soccer. He barely spoke to Ali.

At the end of the trimester, Abdou beat the record in Math, French, Biology, Science, and Classical Arabic.

Madame Petit took him to the principal's office. Mr. Alami handed him an official letter awarding him the scholarship and a contract for his father to sign.

Abdou fought back tears of joy. He hid the documents deep inside his notebooks, terrified that Hsain might find them and destroy them. When he got to Mamma's house, he said nothing. He packed his bag and left for the village.

When his parents heard the news, they thanked Allah. The next morning, Abdou and his father went to the government center in Afourar to notarize the contract.

Abdou had to explain the terms to his father, who couldn't read French. The rules were strict: maintain a high GPA, attend all study halls, stay on school grounds, be in bed by 23:00, wake up by 06:00, and never go on strike.

Then there was the list of required items: five pairs of underwear, soap, shampoo, and dress clothes. Abdou worried his father would refuse to sign because he couldn't afford them. But his father simply took the paper and joined the crowd waiting for the official.

In Morocco, a line always degenerates into a circle. A *Mokhazni* soldier tried to push people back, but occasionally let friends slip to the front. A man in a blue lab coat whispered to new arrivals, taking a few dirhams bribe to bring them closer to the officer.

By 11:30 a.m., the official left for a three-hour lunch. He returned at 3:00 p.m., finally signed the papers, and asked Abdou and his father for their signatures and index fingerprints.

They rushed the signed contract back to the school. An assistant principal congratulated Abdou. "See you in the afternoon on September 17th. Check-in is at 16:00. School starts on the 18th."

Abdou and his father went to visit Addi and Mamma one last time. Abdou's dad thanked them for hosting his son for the past few years, then broke the news: Abdou had earned a scholarship. He would be moving into the boarding school in September.

Father, son, and Ali took a cab back to the village. Abdou's dad was elated, his face lit with a smile. He knew his son finally had a shot at a better life.

CHAPTER 7: Another Summer, Another Coup d'État

When Abdou arrived in the village, he looked for work immediately. He helped his father paint the school and weed the garden.

Uncle Haddou—Abdou's maternal uncle, whose nickname was "The Nation"—came frequently to have lunch and discuss business behind closed doors. Abdou eventually learned that his father and uncle owned some cattle together. They planned to take the herd to the city's open market to sell.

The next day, Abdou woke at 4:00 a.m. to help load three cows and a bull onto a truck. He headed to the city with his father and Uncle "The Nation." With his forced experience as a thief in the market under Hsain's command, Abdou's senses were heightened. He watched everyone approaching his family.

Suddenly, a thief lunged at his father, slashing his chest pocket with a razor blade and grabbing the cash.

Abdou reacted instantly. He kicked the thief like a soccer ball and shouted, "Thief! Thief!"

Uncle "The Nation" jumped on the man, and the crowd joined in, restraining the thief with ropes. Two policemen arrived fifteen minutes later. They took custody of the thief and the stolen money.

"Where is my money?" Abdou's father asked.

"Listen, this money is evidence," the officer said.

"But how can I get it back?"

"You have to be patient. It is evidence."

The police drove away with the thief and the cash, without taking a statement or giving a receipt.

Despite the loss, they sold the rest of the small herd, bought meat and vegetables, and went to Uncle Addi's house for lunch. Afterward, Abdou and his father escorted Uncle "The Nation" to the cab station. The uncle, wise to the dangers of the city, hid his share of the money inside his shoes before departing.

Father and son then went to the police station to claim their "evidence."

No one would help them. Every officer denied the existence of such funds.

They waited for two hours, on edge and ignored. Then, an unmarked car stopped out front. A high-ranking officer exited, flanked by two plainclothes agents. He walked past Abdou's father, then stopped. He stared for a few seconds.

"Are you Uncle Mohammed?" the officer asked.

"Yes, sir."

The officer approached him and kissed his head and hand—a sign of deep respect. "I am the grandson of Haj Moha. I remember you feeding me at school. I will never forget the good you did for all of us students at the elementary school of Timoulilt."

"God save the soul of Haj Moha. This is my son, Abdou," his father said.

"Come inside with me, Uncle."

The officer escorted them into his fancy office, decorated with portraits of the King. Abdou was shaking. He was terrified of the police and the army.

"How are Aunt Fadma and Aunt Itto?" the officer asked warmly.

"They are doing well."

"What brings you here?"

"I sold some cows with my brother-in-law."

"The Nation? How is he?"

"He is good, but a thief stole some of our money. The police took the thief and our money. They said it was for evidence."

The officer's face hardened. He picked up the phone. "Salam. Did we arrest some thieves in the open market today? Who made the arrest? Bring me the paperwork and the detainee."

Five minutes later, a policeman dragged the thief in. The officer stood up, slapped the thief several times, and ordered him back to detention. Then he examined the folder. He looked tense.

He picked up the phone again. "Salam. Bring me the two policemen who arrested the thief."

The two policemen arrived in a hurry, saluting military-style.

"Do you remember this man?" the officer asked, pointing to Abdou's father.

"Yes, sir. He is the victim of a theft."

"Your report here does not say anything about the evidence money."

The policemen shifted uncomfortably. "Sorry, sir, we forgot."

"Where is the money?"

Each policeman pulled 500 Dirhams from his pocket and placed the bills on the desk.

"Uncle, is this the total amount you were swindled of?"

"Yes, son. 1,000 Dirhams."

The police had stolen from the thief, who had stolen from Abdou's father. Without the intervention of the "Angel" from the village, the money would have been gone forever.

The thief was arraigned immediately. The officer ensured the money was returned to Abdou's father on the spot.

"Mr. Mohammed is a poor man from my village," the officer told the judge. "He fed all the children in my school, including me. If Your Honor permits, I must drive him to the village now."

"Why drive him yourself?" the judge asked.

"It is almost 6:00 p.m., Your Honor. There is no transport to the village. Besides, I want to make sure he gets home with his money."

The officer drove Abdou and his father all the way back to Timoulilt.

"Say hello to Aunts Itto and Fadma," he said as he dropped them off.

"Thank you, son. We are proud of you. You are from good blood."

"You too, Uncle. May Allah protect you."

The officer pulled 10 Dirhams from his wallet, gave them to Abdou, and drove back to Beni Mellal.

"Weren't you afraid of the police officer?" Abdou asked.

"No, son. Not of this one. He is one of us. But the other two were rotten eggs. Listen, son—never take anything that is not yours."

"Dad, was he the philosophy student who was in hiding?"

"Yes son, he is reformed now."

Back home, warm tea and crunchy barley bread waited on the table. While Fadma cooked dinner, Abdou helped his father and uncle calculate the profits.

Abdou remembered everything his favorite math teacher, Mr. Bami, taught him.

Profits = Price − Expenses.

It reminded him of fifth-grade math. Once the expenses for the animals, the shepherd and feed were deducted, they split the 1,500 Dirhams profit.

But Abdou's father pushed his share back to his brother-in-law. "You keep this. You will need it. You can pay me once you are at ease."

"Thank you, my dear friend. I will buy a bus and ferry ticket tomorrow to leave Morocco. It is a matter of time before they get to me."

"Abdou," his father said sharply. "Do not say anything about what you heard today. You will put all of us in danger."

Abdou was confused. *Who will be in danger? Who is coming for Uncle?*

The next day, the family gathered at Uncle "The Nation's" home. The mood was somber. Grandmother and Abdou's mother were crying.

"France is not that far," Uncle said. "I will be back soon to take my wife and kids. I won't forget my roots." He turned to Abdou. "Tomorrow, take the Peugeot moped to your cousin Jamal."

"Where are you going, Uncle?"

"To France, son. I am immigrating. Life is hard here. It is a matter of time before they get to me."

"Who?"

"The ones of our time. The ones who control and take everything. Make sure you study hard. Become a doctor or an engineer. Help the family get out of its misery. Promise you will never be corrupt like the ones forcing us to leave "

"Yes, Uncle. I promise."

"Never be obedient or lower your head. Never have a dead heart and take what is not yours."

"Yes, uncle."

Uncle left early the next morning. He made it to Algeciras, Spain, but was returned by border control because he lacked residency papers for France.

Undeterred, he came back to the village and spent a week at the post office—the only place with a telephone. He called every immigrant from Timoulilt living in France, begging them to provide a notarized document stating they would host him. Most refused; they lived in cramped worker dorms or HLMs (housing projects) and couldn't legally host guests.

France parked its immigrants in concrete blocks, isolating them from society.

Finally, Mr. Bouh Salah, the former village nurse, agreed. Salah had no degree, but he had served the poor of Timoulilt for years, injecting penicillin and dressing wounds for free. He sent the notarized document.

Before leaving again, Uncle threw a ceremony for a troupe of Sufi elders called "Foukara.' They chanted and danced like dervishes: "There is no God but God." They prayed for his safe passage. They also drunk boiling water in a ritual the Foukara troupe were known for.

This time, the attempt to cross to France worked.

Weeks later, Abdou picked up a letter at the post office. His father read it aloud to the family:

"My dear friend, Mohamed. Peace be with you. I am in France. Life is tough. I work in a factory filled with smoke. I work every day, but I am making more money than I did in Morocco. Some French workers hate us. They blame everything on the Arabs. I tell them I am not Arab, but to them, we are all foreigners. I live with Salah in a shantytown full of Moroccans, Algerians, and Senegalese. I miss you all. Please take care of my wife, children and mother."

Two weeks after Uncle "The Nation" left, on August 16, 1972, chaos erupted.

Five F-5 fighter jets, piloted by Moroccan airmen, attempted to shoot down King Hassan II's Boeing 727 as he entered Moroccan airspace returning from France.

Another coup d'état had begun.

The jets had taken off from the Kenitra airbase, a facility with deep US ties. Rumors flew that the Americans were involved, though US diplomats furiously denied it.

That night, the family gathered around the shortwave radio, tuning between the BBC, Voice of America, and Moroccan stations, trying to find the truth in the static.

The King had survived. The coup had failed.

"General Oufkir led it," Abdou's father whispered, his face pale. "He is ruthless. He mowed people down from a helicopter in the Rif

Mountains uprising. God saved Morocco from his grip."

The radio confirmed that Oufkir, the King's right hand—the "irreplaceable pillar" of the regime—was dead. Rumors said he was summoned to the palace and shot in the back. Other reports called it suicide.

Repression followed instantly. Within twenty-four hours, pilots were executed, and two hundred officers from the Kenitra base were arrested. Roadblocks appeared everywhere again.

Three days before the *Eid* of Sacrifice, eleven more officers were executed by firing squad. Some shouted "Long live the King" just before they were mowed down.

On the eve of the holiday, King Hassan II delivered a speech that chilled the nation to its bones.

"I am willing to sacrifice one-third of you to rule the other two-thirds."

CHAPTER 8: The Boarding School and the Solidarity of a Good Friend

Abdou spent the rest of the summer of 1972 doing odd jobs. He sold chickens and eggs, but his worst job was harvesting almonds, in his village, for his mother's sister, Aunt Lalla. The tree branches were dry and sharp; they tore his clothes and cut his skin, leading to infections. To make matters worse, the almond husks were covered in a fine, dusty powder that made him itch constantly.

His aunt woke him and his cousin Ali at 4:00 a.m. every day. Their first chore was to fetch water from the stream and water the cows and donkeys. Then they headed to the hills to harvest almonds until 7:00 p.m., only to repeat the water chores at night. They collapsed into bed by 11:00 p.m. to start again before dawn.

One night, his grandmother woke up screaming, "Thieves! Thieves! Oh, neighbors, they are stealing the cows!"

Thieves had taken all the cows and bulls from the courtyard and were about to load them into a large truck. The screams spooked them; they fled, leaving the cattle wandering on the gravel road.

At the end of the harvest, Abdou's aunt gave him a basket of almonds—shells included—to share with Ali. The adult workers were paid several bushels, even though Abdou had worked harder than any of them. But you don't pay family in cash.

The few nuts I received won't even cover the cost of the shirt I tore up harvesting them, Abdou thought bitterly. Ali wasn't happy either.

Despite the exploitation, Abdou managed to save 100 Dirhams, from his chicken business—just enough to register for the fall quarter.

First, he went to the government hospital in Beni Mellal to get the required health certificate. The place was packed. Getting a chest X-ray to scan for tuberculosis took hours. As always in Morocco, the line degenerated into a circle. People pushed and shoved, while a security guard took bribes to usher the morally bankrupt to the front.

After the X-ray, a nurse examined his hair for lice and his skin for signs of disease. When Abdou returned after lunch to pick up his certificate, he read the doctor's note: *"The exam showed signs of malnutrition and scars consistent with physical abuse, but no signs of disease or lice."*

Abdou worried the boarding school would reject him because he couldn't afford all the required items. His father took him to a used clothing market. They bought a toothbrush, toothpaste, and a bar of Palmolive soap. They found two pairs of pants with holes, which his father mended with slightly off-color patches. They bought a used sports coat. "This will do for the suit," his father said.

They didn't buy any underwear.

Instead, on market day, his father bought a couple of yards of black nylon fabric. A neighbor fashioned it into smooth, silky boxer shorts.

That's better than going without, Abdou thought. *It satisfies the requirement and feels good against my skin.*

After lunch, Abdou and his father waited nervously by the main road for a cab. The 4:00 p.m. check-in deadline was approaching fast.

Suddenly, a car slowed down. It was Mr. Bami, Abdou's favorite elementary school teacher.

"Are you going back to school, Abdou?" Mr. Bami asked, leaning out the window. "Hop in. I am going to Beni Mellal to do some shopping. Congratulations on winning a place in the boarding school. You will like it, but you must study hard to keep it!"

"Yes, Mr. Bami," Abdou said.

"Mr. Bami," Abdou's father said quietly, "Abdou does not have all the required items the school asked for."

"Don't worry, Mr. Mohammed. The assistant principal in charge of the boarding school is a friend of mine. I will explain your situation to him. He is a nice guy. But Abdou," he added sternly, "Mr. Bouzgaren demands discipline. Be good."

Mr. Bami stopped at a bookstore and bought Abdou fresh notebooks and pens. When Abdou's father tried to pay him back, the teacher refused.

He dropped them off at the high school and spoke briefly with the assistant principal, smoothing the way.

Abdou said goodbye to his father, who pressed a 10 Dirham note into his hand. "Save this in case you need something. Be good and study hard, son."

Abdou checked in. A supervisor led him to the sleeping quarters—a long, spotless room divided into cubicles. Each section had three walls, a window, no door, and four bunk beds. Abdou stashed his duffel bag in his locker and rushed to the cafeteria to join the other students.

A teacher called the roll and led them to the study hall.

"I will mark you absent, with all the consequences, if you show up a minute late," the teacher warned.

The study hall held about 30 7th- and 8th-graders. Two students were assigned to each table. Abdou was seated next to an 8th grader named Saoud. Saoud had been in all of his previous grades. To his delight, Abdou saw that Ishlef, his friend from Timoulilt, was in the same room.

A piercing alarm sounded for dinner. The students formed a perfect line outside the cafeteria. A teacher called them in groups of eight.

"This is your dining table. Choose a chair; that will be your seat for the rest of the year," the teacher announced. "When the alarm sounds, you have five minutes to get in line. Lunch is at 12:15, dinner at 19:30, breakfast at 06:30. You will always use utensils. One of you will rotate as head of the table weekly. Eat all your food. It is the same food that we teachers eat. After you eat, go to the study hall immediately."

He repeated the speech at every table.

Ikhlef sat at the table behind Abdou. After introducing himself to his tablemates—all 8th graders—Abdou noticed that five of them spoke Moroccan Darija with a heavy Amazigh accent, just like him.

The meal was a revelation. The kitchen staff, dressed in white coats and hats like his father, brought out big bowls of soup, platters of meat with potatoes and sauce, baguettes, and red grapes.

Abdou had never eaten a complete meal like this. With every bite, he silently thanked Madame Petit. He struggled a bit with the knife and fork, but a teacher gently encouraged him. "Don't worry. You will get the hang of it."

Since school hadn't officially started, there was no study hall that night. Abdou and Ikhlef stood outside, laughing and speaking in Tamazight.

"Speak in Arabic!" another kid shouted at them.

Before a fight could start, a teacher clapped his hands. "Lights out early tonight."

Abdou crashed immediately. He slept until the loud siren woke him at 6:00 a.m. He washed his face and brushed his teeth for the first time in his life—he had seen people do it in TV commercials and assumed it was only for the rich. He splashed water on his curly hair, refusing to brush it for fear of ruining the curls.

Breakfast was half a baguette with butter and jam, and a large bowl of café au lait. The coffee wasn't great, but the belly was full.

School started at 8:00 a.m. Abdou's schedule was packed: Mr. Khalil for Civics and Religion, Monsieur Petit for Math, Madame Petit for

Science, Madame Demone for French, Monsieur Hrimi for Classical Arabic, Madame Mamdouh for History/Geography, and Mr. Asenigh for Sports.

The boarding school was a paramilitary institution. Life was planned down to the minute. Order ruled—except when the children of the bourgeoisie acted up. They carried an air of entitlement that grated on Abdou.

One day in the cafeteria, the son of a wealthy family sat in Abdou's designated chair.

"That is my chair," said the entitled brat.

"No, that is my assigned chair!"

"It is not yours anymore. I do not like the sun in my eyes."

A teacher rushed over. Instead of correcting the brat, he yelled at Abdou. "That is his chair from now on. You switch with him."

Abdou sat in the brat's seat, burning with humiliation. At that young age, he realized the truth: rules were made by one class to control another. His pride was wounded, but he complied.

After dinner, Ikhlef met him outside. "I swear, by the soul of my father, I will make his life miserable. Those motherfuckers think they are superior to us. I will take revenge."

"No, Ikhlef. Leave it to Allah. Or wait for the right time."

Ikhlef didn't wait for Allah. Before the next meal, he grabbed a handful of purple berries from the bushes behind the study hall and smashed them onto the brat's stolen chair.

When the brat sat down, his cream-colored pants were ruined. He jumped up, feeling the wet squish. The stain looked exactly like he had soiled himself.

He fled to the dorms to change while the other kids laughed. A teacher accused an innocent student who was laughing too hard and gave him a month of detention, but Ikhlef and Abdou kept their faces straight until they were safe in the study hall.

The rest of the year went well. Abdou got used to the routine of studying and eating well. But as final exams approached, the bourgeois brat returned to Abdou's table in the study hall.

"You will help me with math," he demanded, tossing two Dirhams onto the desk. "From now on, I will sit next to you." He turned to Saoud. "Move to another table. I will sit next to this uncivilized Amazigh."

Saoud didn't flinch. "This is my designated seat. I will not move. Besides, Abdou and I have been in the same classes since 6th grade. And if you say anything else about Amazigh people, I will punish you, you bundle of bones. We are all Moroccans, so shut your trap."

Abdou realized he had a faithful ally. Emboldened, he stood up.

"You will not sit near me. I will not help you. Yes, I am poor. But I am a good student. I won this spot. I bet your rich dad paid someone to get you in. What is your GPA, anyway?"

"Who are you talking to?" the brat sneered.

"My uncle defended this country while your side was with the colonialists," Abdou shot back. "Traitors. I don't want you near me."

Saoud stood up, his eyes flashing with anger. "You will not take my seat, and you will leave Abdou's cafeteria table. I will take your old chair, and Abdou will take his back. I don't like your attitude. Your father may know people, but I know people, too."

Ikhlef and four other students stepped forward, surrounding the table. The solidarity was absolute.

The brat looked around, terrified. He shut his mouth and walked away. From that day on, he sat in Saoud's old chair, and the balance of power shifted.

CHAPTER 9: The Holidays, the Executions, and a Rebellion

The fall trimester ended well for Abdou. But when he returned to the boarding school after the New Year, he was summoned to the principal's office along with Ikhlef, Saoud, and three other kids from his study hall.

Waiting for them were the father and uncle of the mean-spirited bourgeois brat from the previous term.

The principal expressed his disappointment and anger. "You are all in detention until Eid," he barked. "You cannot leave the school grounds. You were mean to your nice classmate."

The kids tried to defend themselves, explaining that the brat was the aggressor, but the brat's father raised his hand.

"I trust my son. He is better than all of you," the brat's father sneered.

It was clear where the boy got his attitude. The apple did not fall far from the tree. The principal sided with the wealthy landowner, a microcosm of the Moroccan system where the "haves" consistently beat the "have-nots."

But Saoud refused to be subdued.

"Your son is not better than any of us, sir. He was mean to all of us. He feels entitled. He thinks he owns the place. He victimized Abdou and tried to victimize me. I witnessed everything, and so did the entire class."

"Who are you to speak to us that way, you poor Amazigh peasant?" the brat's uncle demanded.

"I am the nephew of General Ben Bouhali," Saoud replied calmly, his voice steady. "I am not poor, and I am proud to be an Amazigh. I spoke to you politely, sir. Please don't insult the Amazigh people. We are all Moroccans. Remember God, the Country, and the King."

The room went deadly silent.

"May God save the General's soul," the principal stammered, his face paling. "I did not know he was your uncle. My brother served under him. Both died in the coup d'état. I know your brother, then, son."

"Which one, sir?" Saoud asked coldly. "The head of the *Gendarmerie Royale*, the one in the Rapid Intervention Commandos, or the parachutist?"

The principal ended the meeting immediately. He whispered something to the brat's father, and the adults hurriedly ushered the boys out.

It all depended on who you knew in Morocco. Abdou was glad Saoud was there. The mere mention of his family name paralyzed the influential men who had wanted to humiliate them. The detention was never enforced.

Soon, the *Eid* holiday arrived around January 13, 1973. It should have been a time of celebration, but fear reigned across the country.

More officers involved in the 1972 coup were executed during the sacred holiday.

A repressive era had begun. Group trials of dissidents took place in Marrakech and other cities. At least three of Abdou's village elders—former members of the Liberation Army—fled to Libya to escape the purge. The student union, UNEM, was banned, and the crackdown on opposition parties turned ferocious.

On March 2, 1973, tensions boiled over again. The leftist wing of the National Union of Popular Forces attempted an armed rebellion. Guerrilla fighters planned attacks in the Atlas Mountains to coincide with the anniversary of the King's ascension to the throne.

The rebellion was poorly planned and failed miserably. Civilians and military personnel were rounded up. At least twenty-one people were executed without trial. Even those found not guilty by military courts were often snatched by security forces outside the courtroom and disappeared into secret detention centers.

This marked the beginning of the "Years of Lead"—one of the darkest periods in Moroccan history. Rumors swirled that Muammar Gaddafi was funding subversive camps to destabilize the region, and US intelligence agencies watched closely, terrified that the King's fall would cost them their strategic listening posts in the Cold War.

When Abdou returned to his village during spring break, he found his father in distress—not from politics, but from bureaucracy.

In Morocco, every family required a booklet to record births and deaths. When the government created Abdou's father's registry, a clerk made a catastrophic error: they listed his mother-in-law as his biological mother.

To fix it, he needed witnesses to prove his real lineage. But he had left his village as a boy, and most of the elders who knew him were dead. Correcting the document took over a year of agonizing legal battles.

Simultaneously, a bill arrived for 4,500 Dirhams in back rent—more than seven times his monthly salary. The school where he worked required him to live on the premises as a janitor, cook and night guard. The housing was supposed to be free, just as it was for the teachers. Yet, he was the only one billed. To make matters worse, the government stopped paying his salary entirely while the dispute dragged on.

Abdou watched his father pushed to the brink, bombarded by billing threats while trying to feed his family. He felt helpless and deeply angry at what the government was doing to them.

Abdou channeled his anger into his studies. After eighth grade, he moved with Saoud to a new high school, and dorm, to focus on STEM—Science, Technology, Engineering, and Mathematics.

The new school was rigorous. Abdou excelled in math, physics, and biology, but he soon encountered Mr. Escond, his 10th-grade biology teacher.

Mr. Escond was a French national with a lingering colonial mindset. Despite Abdou ranking number one in the class, Mr. Escond brought him applications to drop out.

"Abdou, there won't be any jobs in Morocco in the future," the teacher said with feigned concern. "Go join a trade school. Become a lab technician."

Abdou discussed it with his father.

"Ignore him," his father said. "He is a racist. He wants to keep you down."

"But Dad, this guy is known to fail good students."

"I told you to ignore him. Study hard. Climb as high as you can. Someday, you might become a professor or a doctor."

A few weeks later, Mr. Escond offered Abdou another application to become a veterinary assistant. Abdou refused. Sadly, some of his less fortunate friends took the bait and dropped out, abandoning their potential.

By 11th grade, Abdou was selected for the Experimental Sciences track. By 12th grade, he was studying philosophy, advanced calculus, and projectile physics. He learned to compute firing angles and velocities. Luckily, he also had a new, fair biology teacher.

The pressure of the looming Baccalaureate exam was immense. The success rate was notoriously low. Abdou formed a study group with Saoud and El Himer. They studied until midnight every day, waking up at 5:00 a.m. to review before class.

The political tension outside the school walls seeped in. Students went on strike to protest the death of Saida, a teacher who had died in police custody under torture or hunger strike. Abdou's philosophy teacher was arrested three times that year, accused of "importing foreign ideas" and corrupting the youth.

A week before the exam, Dr. Dobresco, a Romanian scholar, reviewed calculus and complex numbers with them. Abdou felt ready.

Two days before the exam, he went home to gather his strength. As he walked past the school residence, he was ambushed. The wife of the corrupt elementary principal—the one who had hated his father—jumped out with her daughter. They screamed at him and spat in his face, trying to provoke a fight.

Abdou froze, fists clenched. If he fought back, he would be arrested and miss the exam.

"I know their plan," a neighbor, Mrs. Majdi, whispered as she dragged him away. "They want you arrested so you fail."

"I know."

Abdou swallowed his pride. He wiped the spit from his face and went home. He felt humiliated, but he refused to let them win. He slept for twelve hours. To avoid further trouble, he stayed with his grandmother and uncle for the rest of the weekend.

On Sunday, he returned to the city. Since the boarding school was closed, he stayed with Uncle Addi and Aunt Mamma. Cousin Hsain was there. Hsain was preparing to retake the Baccalaureate for the fifth time. He was training to be a teacher but kept failing. Abdou felt uncomfortable around his childhood tormentor, but he had no choice.

On the days of the exam, Hsain insisted on walking Abdou to the high school.

Abdou felt confident. The final exam was History and Geography. As they walked, they quizzed each other. Abdou asked Hsain about Nazi Germany and Fascist Italy, summarizing the key differences, and similarities.

When the exam started, Abdou turned over the paper and read the

prompt:
   Answer one question in detail:
   1. Compare Nazi Germany with Fascist Italy.
   2. Compare the US economic geography with that of the Soviet Union.

Abdou smiled. He chose the first question. He had written a paper on it and had even read *Mein Kampf* to understand the historical context of 1925. He wrote for two hours straight, filling six long pages of foolscap paper.

"That is a lot of writing, son," the proctor said as Abdou handed it in. "I was watching you; you never lifted your pen. Good luck."

Abdou walked out of the classroom, feeling light as air. Hsain was waiting for him.

"Aren't you glad we discussed history this afternoon?" Abdou asked, smiling, thinking Hsain would be grateful for the review.

Hsain's face twisted in rage.

"You son of a bitch! How did you predict the exam?"

Hsain punched Abdou hard in the chest.

Abdou collapsed, gasping for air. His heart nearly stopped. He felt dizzy and threw up on the pavement. Hsain turned and ran away.

Teachers and the school nurse rushed to help, but the damage was done. Abdou dragged himself to his parents' home, leaving his belongings at Aunt Mamma's. He never wanted to see Hsain again.

A week later, the results were published in the newspapers.

Abdou passed. Hsain failed again.

Droves of women came to Abdou's parents' home, chanting and ululating in celebration. Men came to shake his father's hand.

Abdou immediately began preparing for college entrance exams. He took three: one for the Institute of Agronomy and Veterinary Medicine (IAV), one for the School of Civil Engineering, and one for the Air Force Academy.

He was offered a spot in all three.

Abdou spent countless nights awake, staring at the ceiling, deciding which future to choose.

CHAPTER 10: In Memory of Dean Bekkali, Farmer Khammar, and Professor Pascon

Abdou decided to study at the prestigious *Institut Agronomique et Vétérinaire Hassan II* (IAV).

Saoud and Abdou became roommates again. They shared a spacious, clean dorm room. The student body was the cream of the crop—top students from every region in Morocco and across Africa. The faculty was international, dominated by French, Belgian, and American professors.

At the helm was Mr. Bekkali, the Dean. He was a towering figure, a laureate of a top French engineering school, and credited with building IAV into a world-class institution with the full support of King Hassan II. He constantly reminded his students that failure was not an option.

The program was rigorous, run with paramilitary discipline. Classes ran from 8:00 to 18:00, Monday through Friday, and Saturday mornings until noon. Failing a single class meant expulsion.

But for those who survived, the rewards were immense. The living conditions were excellent. Students received a monthly stipend of 550 Dirhams—twice Abdou's father's salary—plus free room and board. They even had a student-run pub that served coffee, tea, and beer on

Saturdays.

Abdou saved most of his stipend and gave it to his father at the end of the year to rebuild their mud house with brick and cement.

IAV was also a hub of international cooperation. The University of Minnesota maintained an office right in the administration building, managing a USAID grant to train graduate students in the US. Through this "Minnesota Project," American companies such as Cargill and 3M entered the Moroccan market, and top professors from around the world came to teach.

But education wasn't just about theory. In his third year, as part of his field training, Abdou and Saoud spent forty-five days living in a small farming village with a poor farmer named Khammar.

The objective was to analyze complex rural systems across production, economics, and sociology.

Khammar lived in a modest house with his elderly mother, wife, seven children, and his brother's family. His brother had immigrated to France, leaving Khammar to feed a small army.

Khammar was mute, but his intelligence and dignity spoke volumes. He owned four small fields, totaling no more than six acres. He practiced crop rotation religiously: turnips and potatoes in the irrigated field, peas and onions in another, wheat in the third, and fava beans in the fourth. He used manure from his five cows and ten sheep as fertilizer because he couldn't afford chemicals.

"He knows about breeding and animal selection," Saoud noted one day as they watched Khammar take a cow to a neighbor's prize bull. "He's applying genetics without ever setting foot in a classroom."

Despite his poverty, Khammar refused to let the students buy food. Abdou and Saoud had to be sneaky, buying meat and vegetables at the weekly market to contribute to the family pot.

They spent every waking hour with him—plowing, harvesting, selling goods at the market. At night, they wrote their journals and tutored one of Khammar's children, who walked five kilometers to school every day.

During the winter quarter, trouble arrived in a pickup truck.

A prominent landowner with over 300 acres adjacent to Khammar's land showed up with his two adult sons. He started yelling, accusing Khammar of plowing five feet into his property.

"No, Mr. Haj," Khammar signed and gestured to the rock markers. "The line is here."

The landlord's sons didn't listen. They jumped on the mute farmer, knocked him to the ground, and started beating him with sticks.

Abdou felt a volcanic anger erupt inside him. It was the same injustice he had seen all his life.

He and Saoud jumped in, tearing the aggressors off their host.

"Shame on you for beating a man older than your father!" Saoud shouted.

"Don't touch him again, or we will beat the shit out of you!" Abdou added, fists clenched.

"*Yalla*! Let's go to the authorities," one of the sons sneered. They sped off.

Hours later, a jeep of *gendarmes*—the rural police—arrived, escorted by the landlord. They arrested Khammar, threw him in the back, and drove away.

Distraught, Abdou and Saoud hired a local to drive them to the *Caid's* (Sheriff's) bureau. They flashed their IAV student IDs to the *Mokhazni* guard and demanded to see the Sheriff.

"Sir, we are from IAV Hassan II," Saoud declared. "We witnessed the assault. The landlord and his sons beat the farmer. He is the victim. If you do not release him, we will return to Rabat and report that our government-sanctioned training was stopped by local corruption."

"Do not talk about the gendarmes, or you will join Khammar in his cell," the Sheriff warned. Then he paused. "My brother-in-law graduated from IAV. I know how you students can stir the pot. Is Bekkali still the director?"

"Yes, sir. And he has the ear of His Majesty King Hassan II."

The bluff worked. The Sheriff ordered the guard to release Khammar.

"This time, you go free," the Sheriff told the farmer. "These two students saved you. I don't want trouble with Rabat."

Khammar was roughed up but overjoyed. The village rumors spread instantly: Abdou and Saoud were powerful men connected to the Palace. It wasn't true, but the Student Union's power and Director Bekkali's were real enough. From that day on, even the gendarmes gave them rides.

At the end of the spring quarter, IAV bused all the host farmers to Rabat for a weekend celebration.

Khammar arrived wearing a clean white *djellaba* and yellow *babouches*. He had shaved his weathered face. He hugged Abdou and Saoud like his own sons.

"How are you doing, my sons?" he gestured. "I brought a letter from my brother in France. Will you read it for me?"

They took him to the student pub, which was reserved for the farmers—tea and coffee only, out of respect for the rural elders. Abdou opened the letter.

"Dear Brother... Coming from a small village to this big city was not easy. It took me six months to find work. Some of my French coworkers hate us. They spit at us. They say we should go home. But I will lower my head and work hard... Signed, Your Dear Brother."

Tears flowed down Khammar's face.

Later, they took him to the Sociology Department to meet the famous French-Moroccan sociologist Paul Pascon. Pascon, realizing Khammar was mute, treated him with profound respect. "Marhaban a Sidi, Marhaban," he kept saying. *Welcome, sir.*

Pascon often told his students, *"When an old man dies, an entire library burns. Document everything."*

Back at school, the political situation in Morocco was deteriorating.

In his fourth year, Abdou applied for a passport. Instead of a passport, he received a blue summons from the police.

He went to the station. They claimed they didn't send it. He went back to school. The summons came again. And again. Each time, the police played dumb. It was a psychological game. Abdou lived in terror that he would simply disappear, like so many others during the Years of Lead.

Finally, the Director of Academic Studies dragged him to Dean Bekkali's office because he was missing too many classes.

"They keep summoning me, sir," Abdou explained, shaking.

Mr. Bekkali slammed his pipe on the desk. "You will not miss any more classes! Let the police come to you—and to me. You get back to class, son. They should stop playing games with your future. I will deal with it."

"Thank you, sir."

Abdou ignored the next summons. The police never came.

But the harassment didn't stop at the university gates; it followed him home.

During spring break, Abdou returned to his village. His mother hugged him tight. "The gendarmes came," she whispered. "They took your notebooks."

There had been a nationwide strike. The authorities were hunting for political agitators. They had interrogated his father, asking which party his sons belonged to.

"If a party exists in this village, they are members," his father had joked sarcastically. The gendarmes didn't laugh.

Abdou fled to a friend's house in Afourer for the night. The next morning, he met Saoud, whose brother was the head of the *Gendarmerie Royale* in Beni Mellal.

Saoud went inside to talk to his brother. He came out thirty minutes later holding one of Abdou's notebooks.

"They suspect you or your brother wrote a letter threatening the elementary school principal," Saoud explained. "It was signed by 'The Muslim Brotherhood'."

"The Brotherhood? We are liberals!" Abdou said. "We hate fanatics."

"My brother knows. He cleared you."

But one notebook was missing—his thickest one, filled with a year's worth of physics and rocket science notes. It was never returned. Abdou was upset, all lecture notes from his professor Gabaro were stolen.

Years later, Abdou discovered the truth. It wasn't the Brotherhood. It was his cousin Hsain who had attacked the principal's house with rocks one night.

The "evil principal" had been trying to destroy Abdou's family for years. He had tried to force Abdou's father to sign a resignation letter in French, knowing he couldn't read it. He had accused the family of inciting the strike after someone wrote graffiti on the school wall.

The principal eventually fought with everyone in the village—teachers, neighbors, and parents. By the end of the year, he was transferred out.

Peace returned to the village school, but Abdou knew the lesson well: the mighty could crash, but they often took the innocent down with them before they fell.

CHAPTER 11: Goodbye France, Hello Minnesota

By the end of the school year, Abdou had decided to go to grad school in Paris. He assumed he had all summer to get his passport and head to France in September.

On July 14th, 1981, Abdou took his last exam. The professors' council and the dean met, and the undergraduate results were announced. Mr. Phillipe Jouve, Abdou's professor and field training adviser, summoned him.

"Congratulations. You are leaving for Minnesota in less than a week, on July 21st," the professor said.

Abdou was stunned. "But I don't have a passport, and I chose to go to Paris."

"That is not the case, my son. Come with me to the Minnesota Project office."

They walked to the administration building.

"We are sending this young man to study at the University of Minnesota next Tuesday, but he does not have a passport," Professor Jouve announced as they entered.

"Ah! Another one!" said Dr. Johnson, the head of the office. "Why didn't you get your passport? I see that you never missed a class in our English program."

"I applied last year, sir, but for some reason, they have been..."

"Okay. Mr. Habib will help you," Dr. Johnson interrupted.

Dr. Johnson was originally from Roseville, Minnesota. He had been a professor at the university before being assigned to lead the project at IAV in Morocco.

Mr. Habib was the office's main problem solver. He knew every bureaucrat in Rabat. If Minnesota Project materials got stuck at customs, Mr. Habib freed them in minutes. If Americans needed broccoli—a rare vegetable in Morocco at the time—Mr. Habib found it. He drove a US Embassy car with yellow diplomatic plates, which opened doors everywhere.

"What is your name, young man?" Mr. Habib asked.

"Abdou, sir."

"Ah! You have an Amazigh accent. *ManZakin* (How are you?)" Mr. Habib smiled.

Abdou was relieved to find a fellow Amazigh.

"Come see me after lunch. Bring your national ID. 1:30 p.m. here, okay?"

"Yes, sir."

Mr. Habib drove Abdou to the passport office at the governor's headquarters in Rabat. It seemed he knew the governor personally. Mr. Habib cut through the red tape, including the fingerprinting at the police directorate. Abdou received his passport in two hours. That same afternoon, Mr. Habib secured his US J-1 student visa.

Everyone needs a Mr. Habib.

On July 21st, the IAV school bus drove Abdou, fifteen of his classmates, and some faculty to Casablanca's international airport. They boarded a plane to New York.

Even though Abdou had taken extensive English classes, he didn't understand a word when the US Border Control asked for his passport.

These Americans eat their words, Abdou thought. His high school teacher had been a Peace Corps volunteer, but the New York accent was a different beast entirely. He quickly learned that real Americans spoke fast, slang-heavy English.

To make matters worse, Abdou lost his only suitcase. He was left with just the clothes on his back and ten dollars in his pocket.

He boarded the connecting flight to Minneapolis. Upon arrival, he was met by a group of welcoming IAV graduate students who had arrived a year or two earlier. That was the beauty of IAV: alumni were a tight-knit family. Strangers became brothers instantly.

Some students brought Abdou clothes and tennis shoes to tide him over. It took two weeks for his lost suitcase to arrive at his dorm. To his surprise, the tags showed it had traveled through Tel Aviv. He wondered why his luggage, checked from Casablanca to New York, had ended up in Israel. It was a mystery that unsettled him.

Abdou and Saoud were assigned to Comstock Hall on the Minneapolis campus. Abdou enjoyed the dorm—formerly all-women—and quickly made friends in the clean, comfortable facility.

A day after arriving, orientation leaders—mostly young blond women—loaded the new Moroccan students onto a yellow school bus bound for the St. Paul campus. A bearded blond man in his late thirties handed each student a fifty-dollar advance on their scholarship.

Abdou was confused when the man spoke perfect Moroccan Arabic. He assumed he was American, but he turned out to be Mhammed Taya, an IAV faculty member in the Soil Science department.

"You will get the rest of your scholarship—$650—in a week," Taya promised.

The students toured the student center, cafeteria, gym, registration office, and library. Abdou was assigned to the Soil Science department. He met his adviser, Dr. James Swan, a soil physicist working on the impact of tillage on water conservation. Dr. Swan smoked a pipe constantly and spoke some French, having married a French woman. Between the pipe in his mouth and the accent, Abdou found him hard to understand.

The next day, the orientation leaders took the international students to register for the summer ESL (English as a Second Language) program—though for Abdou, it was his fifth language.

That afternoon, Abdou and Saoud went to a store near the dorm to celebrate. They bought a six-pack of something called "Root Beer."

They cracked one open and took a swig. It tasted awful—nothing like the Stork beer back home.

"This American beer is like piss," Saoud said, grimacing.

"Yeah. I miss Stork. Americans must have other beers."

When Abdou returned from registration, he found an older IAV grad student waiting for him. It was Mohamed Sabara, a senior student Abdou had seen in the cafeteria back in Morocco, but never spoken to.

"Hello, friend. You remember me?" Mohamed asked.

"Oh, Si Mohamed, how are you? Looking *zaz* (sharp)."

"Put your backpack in your room and come with me. There are a lot of places you need to know about."

Abdou grabbed Saoud, and the three of them walked across campus to Dinkytown, a student neighborhood with a big grocery store called Meyer's and a Rocky Rococo pizza joint.

Mohamed led them to a basement bar called the Valley Pub, where they ran into other IAV students.

"Let's get a pitcher of beer; it is cheaper," Mohamed suggested.

The waiter approached. "What can I get you?"

"A pitcher of Pabst, please," Mohamed said in impressive English.

"Mohamed, the beer here is terrible," Abdou warned him. "The other day, we bought a beer called Root Beer. It was awful. We couldn't even finish the six-pack. It was like sugar, water, and medicine."

Mohamed laughed hysterically. "You hicks! That wasn't beer. That is soda, like Coca-Cola. It's for kids!"

"Well, it said beer on the can," Saoud defended himself.

"All right, cheers. To a new adventure in America. You have a lot to learn, my children," Mohamed said, raising his glass.

"Who are you calling children? I cannot wait to discover more," Abdou said.

"Watch me, my friends. I will show you how to talk to women."

Mohamed walked confidently to a table where two young women were drinking beer.

"Hi. Can we join you? My friends here need to practice their English. They just got here two days ago from Morocco."

"Morocco? Where is that? What language do they speak?" one woman asked.

"French, Arabic, Tamazight, and Darija."

"Ah, you speak French? I know some French," the other woman said.

"Really? Most people here don't speak a second language," Mohamed said, waving Abdou and Saoud over.

"So, where did you learn French?" Mohamed asked the woman.

She grinned. "I've never taken French. All I know is *Voulez-vous coucher avec moi?*"

The three Moroccans burst into laughter. Abdou was puzzled but delighted. American women, he thought, were definitely easier to talk to than Moroccan women.

Mohamed paid the bill, and they stumbled out of the Valley Pub.

"Your English class starts tomorrow at 8:00," Mohamed said as he walked away. "Get a good night's sleep. And always try to connect with other students who speak English. Practice, practice, my friends."

CHAPTER 12: Hello, University of Minnesota

After Abdou took the English placement test, school began in earnest. Classes ran from 8:00 to 11:30 and 13:00 to 16:00. It took a few weeks to get used to the dorm food—no baguettes, no rich sauces—but the breakfast was excellent.

To Abdou's surprise, he received an additional $1,500 check. It was a "settling-in" allowance, meant to cover rent and furniture once the English program ended.

Abdou met international students from Japan, Turkey, France, Denmark, Finland, Senegal, and Belgium. On Fridays, the ESL program hosted a happy hour at the Minnesota International Center (MIC). Abdou went a few times, but he didn't click with the crowd. Instead, he started going to the Valley Pub with his Moroccan friends.

Then, reality hit. Abdou received a notice that his dorm payment was late.

He was confused. He had assumed the scholarship covered everything directly, as it did at IAV. He didn't know he had to pay the bill himself. He had already spent the $1,500 "settling-in" money. He had only $300 left, and the dorm bill was $400.

He was in deep trouble. Saoud couldn't help; he only had enough to pay his own bill.

Abdou hesitated, but he had no choice. He went to Ron, the program accountant, and asked for an advance.

Ron was perplexed. "How did you spend $1,500 in less than a month?"

"Buying things, sir. Some beer, some books, clothes."

"How much beer? A beer costs $1.25!" Ron argued.

"Sir, I tell them to keep the change."

Ron froze. "How much do you give them?"

"Sometimes $100, sometimes $20. The waiters were always very nice and happy to see me."

"Abdou! Abdou!" Ron shouted, burying his face in his hands. "The damn beer is $1.25! You cannot leave those kinds of tips. Even billionaires don't do that!"

"But sir, I saw that in American movies. They always say, 'Keep the change.'"

"You are not in a freaking movie! This is real life. Besides, a dollar is not a Dirham. A dollar is worth six Dirhams. You spent six times the monthly salary of a Moroccan engineer on tips! Are you on drugs?"

"No, sir. No drugs. Me, no drugs. But I need money for the dorm."

Ron sighed, agitated but sympathetic. He pulled out his personal checkbook and wrote a $250 check.

"This is my own money," Ron said sternly. "Not the University's. Not Morocco's. You bring me the $250 at the end of the month when you get your stipend. And never leave more than a dollar as a tip. No more 'keep the change' horse shit. Be smart. Understood?"

"Yes, sir. Thank you, Mr. Ron."

Ron had saved his neck.

Abdou took the bus from St. Paul to Minneapolis and paid the dorm immediately. Humbled, he hunkered down. He changed his routine. He ate every meal at the dorm, studied in the library, and listened to English tapes. When his stipend arrived, he paid Ron back immediately. He still went to the Valley Pub on Fridays, but the "keep the change" habit was dead.

In his two weeks of generosity, Abdou must have tipped over $1,000. It explained why the staff stopped checking his ID. When the line was out the door, the bouncer would yell, "Hey, you, Moroccan! Get in. We already checked you."

At the end of July, the ESL students were sent to live with American host families for cultural immersion. One of Abdou's classmates stayed with Bud Grant, the legendary coach of the Minnesota Vikings.

Abdou was paired with an African American family. His host mother, Betty, was a former teacher who now worked in healthcare at the University, coordinating programs in Kenya and Nigeria. She was raising four kids on her own. She was a generous women.

Betty picked Abdou up on a Friday. She stopped at a restaurant, ordered a large pizza, and—to Abdou's confusion—packed the leftovers in a box to take home. In Morocco, no one took leftovers home.

They drove north on I-94, passing through a tunnel. Abdou saw signs for Brooklyn Center.

The next morning, breakfast was eggs, toast, and coffee. Betty didn't serve pork. Kenneth, her seventeen-year-old son, took Abdou for a walk.

The neighborhood was rough. Many houses were boarded up, and the grass was overgrown, though Betty's house was well-kept. The local grocery store had no windows; it was boarded with plywood and had a reinforced door. Inside, there were cigarettes, soda, and junk food, but nothing nutritious. The nearest store with fresh vegetables was thirty minutes away.

Before arriving, Abdou had watched the miniseries *Roots*. Walking through this neighborhood, seeing the graffiti, he thought about the economic disparities he had read about.

"That's gang graffiti," Kenneth explained. "Don't wear blue or red. You don't want them to confuse you with a target."

Two girls shouted at Kenneth from a porch. "K-Man! Who is the guy?"

"He is from Africa."

"How come he is not black?" one girl asked.

"Not all people in Africa are black. He is from Morocco."

"Can you bring him over? I want to show him my couch. I like his curls," she teased.

Kenneth laughed.

"What is going on?" Abdou asked.

"She likes you."

"But... will her parents allow that?"

"Maybe."

After lunch, Kenneth taught Abdou how to use the bus system. They took the bus to downtown Minneapolis, transferred, and rode to the University. Abdou was impressed. The buses ran on a schedule, unlike the chaotic, crowded buses in Rabat.

Sunday morning, the family went to church. It was twenty minutes away, and the congregation was entirely African American. Abdou felt the weight of segregation he had only read about.

The church was lively. Men wore suits; women wore their best dresses. It was Abdou's first time in a church. When the pastor raised his hands and shouted, his voice rising in agitation, Abdou was scared. He thought the man was yelling at him. But it was just the style of worship. The pastor's robe reminded him of Desmond Tutu.

Everyone sang, "Praise the Lord!" The heat and humidity inside the building were intense; one worshiper fainted, overcome by the spirit and the lack of air conditioning.

Abdou tried to fit in. When they stood, he stood. When they kneeled, he kneeled.

Then came the collection. A woman passed a basket full of money to Abdou. Confused, he said, "Thank you," and set the basket on his lap, thinking it was a gift.

Betty quickly grabbed the basket, put five dollars in, and passed it along. Abdou burned with embarrassment.

Later, he stood in line for communion—a piece of bread and a sip of wine. Betty gently pulled him back, explaining that it was only for Christians.

One Saturday, Abdou asked Kenneth to take him shopping. They cut through neighborhoods and woods until they reached a massive barrier: Interstate 94.

"We have to be careful," Kenneth said. "Make sure the cars are far enough away."

They ran. They sprinted across the freeway in a few giant steps, dodging traffic moving at 60 miles per hour.

Abdou bought a pair of Wrangler jeans. On the way back, they crossed the freeway again. This time, drivers slammed on their brakes. Tires screeched. People rolled down their windows, giving them the middle finger and screaming, "Assholes!"

"Run!" Kenneth yelled. "Before the police come!"

Abdou ran, clutching his new jeans, unaware he had just committed a crime.

In September, the English program ended. Abdou passed the Michigan English Language test and registered for the fall quarter. He moved into a one-bedroom apartment in Dinkytown with a classmate from IAV. Both of them would move to Nebraska for the spring quarter.

Winter arrived early. In November, the snow began to fall. Abdou's boots were too big, and he slipped and fell constantly until he learned the "penguin walk" necessary to survive a Minnesota winter.

On Thanksgiving, Betty drove him to her daughter's house. Before leaving, she gave him a heavy-duty down jacket.

"This is for the real cold," she said. "Your light jacket won't do."

She was right. Abdou shivered in that jacket many times while waiting for the Route 13 bus between campuses.

The fall quarter went well. Abdou earned an "A" in all his graduate classes: Technical Writing, Soil Physics, and Statistics. But in December, his roommate moved to Nebraska. Abdou was left with a $350 rent bill he couldn't afford on his own.

On Christmas Day, Betty picked him up again. The party was full of international students from Brazil, Kenya, Nigeria, and France. Everyone was dressed formally except Abdou, who wore his University of Minnesota sweatshirt and snow boots.

On the drive back, Abdou was quiet. He kept thinking about the poverty he had seen in North Minneapolis and the segregation in the church. He tapped the dashboard in frustration.

"Are you okay? "Betty asked.

"I am just upset. My mind is wandering between here and Morocco."

"Why?"

"I see segregation here, just like the social ladders in Morocco," Abdou said. "I cross the bridge over the Mississippi, and I see homeless people sleeping in the cold. Isn't this the wealthiest country in the world?"

"Well, I saw many poor people in Africa, too," Betty noted.

"Yes, but Africa is poor because it was pillaged," Abdou said passionately. "The colonizers left, but they forced contracts on us to keep extracting wealth. It's the same system."

"Where did you learn all of this?"

"In our economics classes at IAV. And from history. Look at Patrice Lumumba in the Congo. Killed in 1961 by Mobutu, with help from Belgium and the US."

Betty pulled up to his apartment. She looked at him with a mix of pride and concern.

"Be careful, son," she said softly. "You know too much. Keep studying. Keep learning."

Abdou didn't understand the warning. To him, knowledge was the only way out.

A few days after Christmas, Abdou registered for the winter quarter. He didn't know it yet, but his life was about to turn upside down—so much so that the pain would almost drive him back to Morocco.

CHAPTER 13: Goodbye, My Dear Friend

A few days after Christmas, Abdou, Saoud, and a few friends met at the Valley Pub to plan a ski trip to Wisconsin. They decided to take two cars the next morning.

But when Abdou got home, he found an urgent letter from the St. Paul campus library. He had failed to return a book checked out weeks before finals. The fine was $15, and penalties were accruing daily. He had to resolve it immediately. Abdou called his friends and told them he couldn't go skiing.

On the morning of December 30th, Abdou went to campus, paid his fines, and returned the book.

At 5:00 p.m., he received a call from Saoud. "We're on our way back from Wisconsin," Saoud said cheerfully. "Meet us for pizza and beer around 7:30."

Abdou waited at the pub until past 9:00 p.m. His friends never showed up. Worried, he went back to his apartment and called another friend who had been on the trip in a separate car.

"I got back around 7:00," the friend said. "We all left at the same time."

Snow was falling heavily. Abdou called his friends' apartments repeatedly. No answer.

Around midnight, his phone rang. It was Dr. Purvis, the British director of the Morocco-Minnesota Project.

"Do you live in the same building as Saoud, Elmotad, and Yassine?" he asked gravely.

"Yes. I live near them."

"There has been an accident."

Abdou's stomach dropped. "Are they okay?"

"One is in critical condition. Two are badly hurt."

"Where are they, sir?"

"Regents Hospital in St. Paul."

Abdou called the hospital immediately. The news shattered him.

Saoud was dead.

Elmotad had a broken leg and facial trauma. Yassine had a fractured arm and jaw.

Abdou was devastated. He couldn't stop crying. He spent the night calling Morocco, informing his own family and Saoud's parents, though IAV had already broken the news to them.

The funeral took place the next day. Abdou packed Saoud's belongings and sent them with his body back to Morocco. Every piece of clothing he touched made him weep.

Severe depression set in. Abdou could barely function, let alone study. He spent the next few weeks caring for Elmotad and feeding Yassine, who couldn't eat on his own. The university provided no assistance; Abdou was their only lifeline.

Despite his grief, Abdou had to attend classes. He took Advanced Statistics, Climatology, and Advanced Topics in Soil Physics. Statistics and Soil Physics were manageable, but Climatology was a nightmare. The professor, Dr. Baker, was a World War II veteran who marched into class, dropped his heavy jacket, rubbed his hands together, and launched into dense lectures immediately.

"Compute the short and long waves... detail the energy budget balance under different albedo scenarios."

Abdou struggled to understand Dr. Baker's rapid-fire delivery. He missed the announcement for the first quiz entirely and failed it.

Desperate, Abdou started sitting next to an older student named Jim, a Vietnam veteran who worked for the state climatology office. Jim helped Abdou catch up.

But the depression persisted. Abdou wanted to drop out and go home. His adviser urged him to see a doctor, who prescribed medication and referred him to a psychologist. The therapy focused on grief and survivor's guilt—Abdou felt that Saoud had died in his place. Abdou always sat in the front passenger seat, but on that horrible day, Saoud had taken that spot. El Motad drove the car.

The counseling helped, but the medication made him drowsy, making his 8:00 a.m. Climatology class a struggle. He finished the winter quarter with A's in two classes and a C in Climatology.

Spring quarter arrived. Abdou registered for Plant Physiology, System Simulations, Soil Conservation, and a research paper on dryland farming. He loved the coursework, except for the 8:00 a.m. physiology lectures.

Around April, Abdou went to a party at a friend's house. The room was hot and crowded, so he stepped outside for fresh air. He was joined by a beautiful young woman named Marielle. They walked around the block, talking easily. She was a sociology student who drove a beat-up red car with a hole in the passenger floorboard so large you could see the road rushing by.

They started dating. Marielle invited Abdou to a family picnic at Lake Harriet. He met her family but left a bad impression by eating peas with his fingers—a Moroccan habit hard to break.

A couple of weeks before the end of the quarter, Abdou received a letter: he was being sent to the University of Nebraska for the summer. He wasn't happy about leaving Marielle, but it was a mandatory part of his program.

On a Saturday in June, Marielle drove him to the airport at 6:00 a.m. She wore a beautiful red dress. Everything about her was perfect.

"Call me when you get there," she said, kissing him goodbye.

Abdou's plane landed in Cheyenne, Wyoming. He transferred to a tiny ten-seater aircraft for the flight to Scottsbluff, Nebraska. The turbulence was terrifying; the plane shook like an unbalanced washing machine.

Charlie Fenster, a renowned dryland farming expert, met him at the airport. He looked the part: cowboy hat, checkered shirt, and cowboy boots.

"Howdy, I'm Charlie. How was the flight?"

"Nice to meet you, Mr. Fenster. The flight was scary."

"Call me Charlie. We'll be seeing each other every day until the fall."

Charlie dropped Abdou at the experimental station's residence, five miles from the research site. A party was in full swing.

Wayne, a PhD student, introduced himself in perfect French. "Welcome to the middle of nowhere, USA, my Moroccan friend." He handed Abdou a can of Pabst Blue Ribbon.

"Your French is good," Abdou noted.

"I was a Peace Corps volunteer in Senegal and Mali."

"Ah, section SA20 of the embassy?" Abdou joked, referencing the CIA.

"I cannot confirm nor deny," Wayne laughed.

Shihiro, a student from Japan, waved him over. "Come join us. We have crackers and cheese. This is our cheap happy hour."

Wayne was right: they were in the middle of nowhere. The landscape was nothing but cornfields and Intercontinental Ballistic Missiles (ICBMs) buried in silos deep underground.

The next morning, the group went biking. Abdou borrowed a bike and joined them on the country roads.

Work began on Monday. Charlie gave Abdou a tour of the station. In the afternoon, Charlie's assistant took Abdou to the hospital to get a radiation dosimeter, a required gear when working with a neutron probe.

Abdou settled into a routine: mornings spent sampling soil moisture in the scorching heat, afternoons in the library or analyzing data.

He bought a used bike for fifteen dollars to get around town. The seat was falling apart, so someone had polished it with brown shoe polish to make it look new. In the sweltering Nebraska heat, the polish melted, leaving a permanent brown stain on the seat of Abdou's jeans. It looked exactly like he had soiled himself. He didn't care; his clothes were already covered in grease and oil from the tractors.

Abdou attended a conference in Las Cruces, New Mexico, where he saw a computer interfaced with a spectrophotometer. The technology fascinated him.

Back in Nebraska, Charlie took Abdou to the North Platte research station. Abdou met Berrada, another Moroccan PhD student, and Dr. Garity, a renowned water-stress professor. The fields were littered with Minuteman nuclear missile silos controlled by Warren Air Force Base.

Abdou realized he was standing on a primary target for Soviet missiles at the height of the Cold War.

In early August, Marielle came to visit. They stayed in a rooming house in downtown Scottsbluff. They tried to cook in the kitchenette until the manager scolded them. Broke and hungry, they walked around town looking for food.

They tried to rent a car from "Ugly Duckling Car Rental," but the owner refused them because they lacked a credit card or the $500 deposit. Frustrated, they asked a local man for a restaurant recommendation.

"I don't know if there is a Mexican restaurant around," the man sneered, assuming they only wanted Mexican food.

They ignored the racism and found a pizza place instead.

Marielle returned to Minnesota, and a few weeks later, Abdou finished his training. He joined her briefly in Minneapolis before flying back to Morocco to begin his research in Settat, a dry region south of Casablanca.

He stopped in Settat to meet Dr. Watts, the head of the American project there. Dr. Watts agreed to fund Abdou's research if he changed his experiments to study different water conservation factors. Abdou decided, even though he hadn't cleared it with his IAV adviser, Professor Ouattar.

The research station in Settat was in a desolate area, though thankfully free of nuclear missiles. The Americans lived in the wealthiest neighborhood, which locals called "Dallas" after the popular TV show, because of the massive American 4x4 trucks and the luxurious villas.

Abdou moved into the first floor of a converted villa-lab with Oubaha Lahcen, a project employee and IAV graduate. Lahcen was kind and welcoming. They would become best friends for life.

CHAPTER 14: The Experimental Station

Abdou was fortunate to be surrounded by fantastic mentors. His primary thesis adviser, Dr. Ouattar, was based in Rabat—a brilliant agronomist trained in France and the US. His co-adviser, Dr. Watts, was a soil physicist and engineer from the University of Nebraska, based in Settat.

Because Dr. Watts was on-site, Abdou spent most of his time at the experimental station in Settat rather than at the IAV campus in Rabat.

Although the thesis requirement was a single field experiment, Abdou—ambitious to a fault—designed three. When he presented his plan to the IAV faculty, Dr. Tayeb joked, "You'll need a horse to handle that amount of work."

"It is worth it," Abdou replied confidently. "And I secured the funding."

Later, exhausted and covered in dust, Abdou realized he had vastly underestimated the colossal effort required.

Dr. Watts financed everything. He flew in a hydraulic probe and a neutron probe from the US within a week. Abdou and the project accountant traveled to Casablanca to buy supplies: paper, plastic bags, markers, notebooks, and 250 aluminum tubes, each over two meters long. They even bought a John Deere tractor, delivered the same day.

Abdou learned to operate and maintain the probes in Nebraska, so he drove the tractor himself and managed a crew of five workers. He drilled deep holes, driving the aluminum tubes two meters into the ground. To cap the tubes, he used empty Coca-Cola and Miller beer cans collected from the American researchers, who bought their supplies exclusively from the US Embassy store.

"I'm investing in your research," Dr. Watts told him. "I'm putting you in charge of this equipment. Charlie Fenster highly recommended you."

"I enjoyed working with Charlie," Abdou said.

"Good. I don't want you sitting on your butt at a desk all day like some of the engineers here," Watts warned.

"Don't worry. I prefer the field."

Abdou knew the Nebraska-Morocco project suffered from a lack of data due to drought and low morale. He was determined to change that.

He commuted, fourteen kilometers, between the research station, and Settat. The small town was chaotic—a mix of sixteenth-century rural life and modern shantytowns. Prostitution was rampant; women in modern dresses catered to men with cars, while those in traditional *djellabas* walked the streets for blue-collar workers.

Living between Rabat and Settat drained Abdou's finances. The IAV accountant stopped paying his travel stipend, leaving him over

3,000 Dirhams short. Abdou survived on tea and bread, occasionally treating himself to *Vache Qui Rit* cheese or yogurt. His weight dropped to 125 pounds.

One sunny day, Abdou arrived at the station to find his tractor and hydraulic probe missing. He ran to the field and found a French professor, Taghat, trying to use them. She clearly didn't know what she was doing.

Abdou was furious. He took the tractor back immediately. Taghat complained to the station chief, but the chief sided with Abdou. "Dr. Watts bought that equipment for him," he explained. Taghat stormed off to Rabat, vowing revenge.

As temperatures soared, the work became grueling. Abdou survived a near-fatal motorcycle accident commuting to the station, so he decided to move in. He cleared out a storage room filled with straw, laid down blankets, and built a desk from planks and oil drums. The room was infested with ants.

A warehouse worker took pity on him and offered to share his cleaner room. Abdou gratefully accepted, paying ten Dirhams a day for groceries. They lived on lentils and fava beans, feasting on a meat tagine once a week.

The station manager called Abdou into his office. "That is not a good place for you. As engineers, we don't mix with blue-collar workers."

"Sir, my father is a blue-collar worker," Abdou replied firmly. "I almost got killed commuting on my motorcycle. I need to be here early to work before the heat damages the neutron probe. Please don't make a big deal out of this."

The chief smiled. "You are a real product of IAV and America, aren't you? Fine. You are doing a great job. The institute hopes to hire you later. But... can you stop wearing shorts? I have received complaints. Some people consider it indecent."

"The fanatics complained?" Abdou sighed. "Fine. Even if it's 115 degrees, I'll wear pants. Soon they will force all women to wear burkas."

To avoid the heat, Abdou split his workday: 6:00 to 11:00 a.m., then 4:00 to 7:00 p.m.

But the station chief's assistant fired Abdou's best worker for "leaving early," even though Abdou had put in 50-hour weeks with his workers.

Abdou confronted the assistant. "I changed the schedule for safety! The neutron probe's circuits melt at 110 degrees. I need him."

"That is not my problem," the assistant sneered.

Abdou was exhausted by the pettiness. Then he noticed twenty aluminum access tubes were missing from the warehouse. With help from his roommate, who watched everything, Abdou learned the assistant had hosted a party the night before.

"I believe your brother-in-law walked out with four people carrying tubes," Abdou told the assistant. "I hold you responsible for stealing US and Moroccan government property. Bring them back and reinstate the worker, or I report this to Dr. Watt and the gendarmes."

The assistant folded. The worker was reinstated, and the tubes reappeared.

After finishing his sampling, Abdou went to Rabat to meet Dr. Ouattar. He was thrilled to find letters from Marielle waiting for him, though he suspected some had been stolen or opened.

In his dorm room, he found Naser, his friend and a brilliant scientist at the National Institute of Research (NIR). Naser hadn't been paid in seven months—a common bureaucratic torture.

"You are ugly, broke, and miserable," they told each other simultaneously, laughing and hugging.

Abdou's cousin Ali was also in Rabat, learning English at the American Language Center before a trip to Utah for training. Ali was stingy but street-smart. Abdou and Naser decided to play a joke on him.

"Cousin Ali, I will miss you before you die trying to reach America," Abdou said solemnly.

"What are you talking about?"

"America is far. Your ticket says you fly to Utah. That takes days. If you don't bring food and water, you will starve. They locked me in a windowless room for four days without water."

"You are messing with me," Ali said.

"No," Naser interjected with a straight face. "He is right. And American planes don't have toilet paper."

Ali panicked. The next day, he bought a massive box of supplies: toilet paper, hard-boiled eggs, baguettes, water, and juice. He brought it to Abdou's room to show he was prepared.

Abdou and Naser burst into laughter, slapping Ali on the back. "We got you, cousin!"

Ali left for Utah—without the box—vowing revenge.

Back at the experimental station, Abdou was working in the field when a convoy of American SUVs approached.

He looked up to see Charlie Fenster, wearing dress pants, a white shirt, and a tie, walking toward him with his wife Eunice and the entire Nebraska team.

"Hi, Charlie! Hi Eunice!" Abdou shouted, shaking his hands.

"What are you working on, young man?" Charlie asked.

"Taking bulk density samples."

Charlie nodded, then spotted the tractor. Without hesitation, he crawled underneath it. He emerged moments later, covered in oil and dirt, his pristine white shirt ruined.

"Found a leak in the hose," Charlie announced, wiping his hands.

Eunice didn't blink. She simply led Charlie behind an SUV and handed him a fresh shirt and a pair of pants. She knew her husband well. She was always prepared.

CHAPTER 15: The Thesis Defense

Abdou arrived at the experimental station in Settat, but as he stepped off the bus, he didn't recognize where he was.

The dusty, chaotic road was gone. The area was spotless. Large orange and palm trees lined the boulevard. Moroccan flags waved from every pole. Banners proclaimed *"Bienvenue!"* The place was crawling with soldiers and *gendarmes*.

Abdou felt a knot of unease. He crossed the railroad tracks and tried to enter the station, but two soldiers stopped him at the gate. He showed his National ID and student researcher credentials. After a tense interrogation, they let him pass.

Inside, the station was swarming with armed guards.

"What is going on?" Abdou asked his roommate.

"I don't know. Maybe the King is visiting."

Later that day, King Hassan II and French President François Mitterrand arrived on a royal train. They stopped at the station, had lunch on board, and left. Perhaps the King wanted to show Mitterrand the fruits of Moroccan-American collaboration. To this day, no one knows why they visited the middle of nowhere.

A few weeks later, all the orange trees died. The locals pulled them out of the ground and used them as firewood. They had been planted solely for the show—rootless props for a royal play. It happened often in Morocco: before a dignitary arrived, garbage vanished, trees appeared, and beggars were swept away. Once the VIPs left, the garbage returned, the trees died, and the wild dogs reclaimed the streets.

Abdou finished his data collection and returned to Rabat. He was nervous. He had to finish his analysis before IAV closed for the summer on July 30th.

He secured a room in the Agricultural Engineering department to enter his data onto punch cards. But a technician kicked him out.

"This machine is only for faculty," the man sneered.

"But I have permission from the department head!"

The technician didn't care. Abdou learned early that in Morocco, mediocre people gained power by stripping others of their rights.

To make matters worse, as he walked across campus, he tripped. His box of punch cards was scattered across the pavement. The wind caught them. Eight hours of data entry vanished in seconds.

Abdou sat in the cafeteria, defeated, reading a letter from Marielle. She was coming to visit in September. That thought gave him the strength to push on. He had to defend his thesis in October so they could be together.

He remembered a trick he learned helping his father: making master keys. He went to downtown Rabat, bought blank keys and a file, and spent hours smoothing the grooves until he had a set that could open almost anything.

At night, Abdou broke into the Agricultural Engineering department. He slipped into the computer room and worked from 10:00 p.m. to 6:30 a.m., entering data in the silence. He slept from 7:00 to 10:00 a.m., then returned to another computer room to analyze his datasets.

The lab used Honeywell Bull mainframes. The manager, Mr. Taws, was a brilliant statistician who took a liking to Abdou. They bonded over advanced statistical methods such as the Newman-Keuls and Generalized Linear Models.

Taws' assistant, however, was less kind. She limited Abdou to four hours a day, even when the lab was empty. But Taws had Abdou's back. He returned a textbook Abdou had lent him, and inside was a lab key.

"Use this at night," Taws whispered.

Abdou worked around the clock. By July 29th, he had two suitcases full of computer printouts. He retreated to his village, Timoulilt, to write. He secured a classroom in his father's school and worked tirelessly, crunching the final numbers on his Texas Instruments TI-59 calculator.

At the end of September, he handed the final draft to Dr. Ouattar.

While his adviser reviewed it, Abdou took three days to relax on the beach, mentally preparing for his defense.

Dr. Ouattar approved the draft. Now came the hurdle of printing. The school had stopped Abdou's stipend in July. He was broke. He couldn't afford a typist or printing costs. But his friend Naser, who had finally received eight months of back pay, lent him 2,000 Dirhams.

Abdou hired a secretary who typed 70 words per minute. He proofread furiously, retyping pages himself when needed. He printed copies and personally delivered them to his jury members.

One jury member, Dr. Fred Troeh from Iowa State University, asked him a critical question.

"Did you make slides yet?"

"No, sir. I will use transparencies."

"Don't do that. You have nice data. I will help you."

That weekend, Dr. Troeh converted his hotel bathroom in Rabat into a darkroom. He developed the film and helped Abdou create thirty professional slides.

The defense was a triumph. The jury—comprising Moroccan, American, and Canadian scientists—awarded him "High Distinction." The report concluded that Abdou would make a great scientist.

But the victory was bittersweet. IAV did not hire him. The National Institute of Research (NIR) offered him a job in Settat, but Abdou knew how poorly they treated scientists like Naser. He rejected it, pinning his hopes on ETAGOS, a large state-owned company whose CEO had a US degree.

Marielle arrived on the day of *Eid*. She came with her friend Christine, who was married to Abdou's friend Mo. The four of them stayed in Mo's apartment in Rabat before heading to the village.

The trip to Timoulilt was a disaster. The six-hour bus ride took eight. The bus stopped constantly to cram in more passengers. Marielle was distressed by the unsanitary bathrooms at the stops—holes in the ground with no toilet paper.

Abdou's family didn't know he was coming, let alone that he was bringing an unmarried American woman.

They arrived at 6:00 p.m.

"Salam. This is Marielle. She is from America," Abdou announced.

"Salam, my daughter," his mother said, touching Marielle's hair gently. "Welcome. Come in for tea."

His mother spoke only Tamazight. His father spoke broken French.

"*Vous parlez Français?*" his father asked.

"*Un peu, Monsieur,*" Marielle replied, looking overwhelmed.

They squeezed into the tiny house at the school. The next day, Abdou took Marielle to meet Aunt Lalla.

"She is young and beautiful," Aunt Lalla declared loudly. "And not with oversized breasts. I like her already."

"Lalla! That is enough!" Abdou hissed, refusing to translate.

"Did you graduate?"

"Yes, Lalla. I am done."

Lalla launched into a piercing ululation of joy. "Finally! Your parents will harvest the fruit of their sweat." She knew the hell Abdou's parents went through to raise their children.

Abdou showed Marielle the village—his old adobe house, the grandparents' home.

"You didn't just travel miles," Marielle whispered, looking at the women carrying water jugs on their backs. "You traveled centuries."

She saw women walking in negligées without hijabs, and others with facial tattoos resembling nuns.

At the market, everyone stared. Whispers followed them like a wave.

"Who is he?"

"The son of Fadma and Mohamed. She is a Christian. *Mirikania*."

"What are they saying?" Marielle asked.

"*Therguig*," Abdou laughed. "Gossip."

Kids followed them, hand out. *"Un peu d'argent, s'il vous plaît."* They thought they were French tourists.

"Salam, whose kids are you?" Abdou asked in sharp Tamazight.

The kids scattered, embarrassed to realize the "tourist" was one of their own.

CHAPTER 16: Three Jobs in a Month

Abdou received a job offer letter from ETAGOS. His parents were thrilled; ETAGOS was a large state-owned company known for paying well and offering excellent benefits.

The letter listed a mountain of required documents: certified copies of his birth certificate, national ID, degrees, and residence permit, plus a judicial record, a police report, and three photos. He had to present everything in Rabat on November 15th at 8:00 a.m.

Abdou left early in the morning, walking six kilometers to the government center in Afourer. He managed to gather the documents and walked back, arriving home after sunset.

He found Marielle sitting on a stool in the middle of the room, her feet dangling. His sisters were applying henna to her hands and feet, so she couldn't touch anything until it dried.

There was a problem: Marielle was allergic to cats, and Abdou's house was full of them. Her nose was running uncontrollably. Since she couldn't use her hands, one of Abdou's sisters was wiping her nose for her. Marielle looked mortified, especially when a large blob emerged.

"How did you manage without me translating?" Abdou asked, trying not to laugh.

"Not bad. But I ate something terrible for lunch."

"What was it?"

"It felt like chewing on an ear and tasted like fried blood."

"Ah! You ate the Moroccan *kourdas*—the grenade!" Abdou laughed. "It's made from dried lungs wrapped in sheep stomach and intestines. It's salted and sun-dried for months. It looks just like a hand grenade."

"Ugh!" she groaned.

The next day, they traveled to the court in Beni Mellal to get Abdou's judicial record, proving he had no criminal judgments. Then to the police station for the anthropometric card—a surveillance report proving he had no outstanding warrants.

They stopped at Cousin Ali's office. Ali took them to a coffee shop for a *Nouss-Nouss* (half espresso, half frothed milk) before taking them home for lunch.

They met Uncle Addi, Aunt Mamma, and Cousin Hsain, who was visiting with his wife. Addi and Mamma were living better now that their children were working. Ali proudly announced that Abdou was a "big shot" who had landed a job with ETAGOS.

Hsain invited them to visit him in the small mountain city where he taught high school physics. Abdou tried to decline, claiming he was swamped, but Mamma and Addi insisted.

"I expect you there on Wednesday," Hsain said. "We have a naming ceremony for my daughter. Our friend Said, the taxi driver, will bring you."

"Okay," Marielle agreed quickly, desperate to escape the cats in Abdou's house. She didn't know Hsain's character yet.

After lunch, they said goodbye to Ali. They tried to return to the village, but the family insisted they stay overnight. They visited the *Ain Asserdoune* springs and the ancient fortress overlooking the city.

One problem Marielle faced was random women asking her questions on the street. When she replied, "Makaynch Arabia" (No Arabic), they assumed she was a *fils à papa*—a rich, stuck-up Moroccan who refused to speak the local language. With her brown hair, she blended in too well.

On Wednesday, they visited Hsain. Abdou was apprehensive but went anyway, out of respect for Uncle Addi. At the high school where Hsain taught, Abdou met three American Peace Corps volunteers. Among them was a young man named Chris Stevens, who would later become the US Ambassador to Libya and tragically die in Benghazi.

A few days after the ceremony, Abdou, Hsain, and Marielle went for a walk. Hsain saw some women walking in a plowed field.

"Hey! Don't walk through my field!" Hsain shouted at the women.

"Why? What did you plant here?" one woman shouted back.

"I planted *Ibella*!" Hsain shouted. It was a crude term for penises.

The women screamed at him. Then they recognized him—they were his students.

Hsain quickly detoured to avoid them, admitting he hadn't recognized them from a distance. He was nearly forty, yet still unbalanced.

After a week, Abdou and Marielle packed their bags. Hsain handed Abdou two checks totaling 3,000 Dirhams.

"Cousin, I know you are broke and your parents don't have money. Take this to help you settle into your new job."

Abdou was surprised and overjoyed. He thanked Hsain profusely, though a small part of him remained suspicious.

A few days later, Abdou tried to cash the checks at the bank. The teller looked at him grimly. "These checks were reported stolen. The owner canceled them."

Abdou left the bank in tears. He told his parents and aunt what happened. Hsain hadn't changed. He was still the same cruel tormentor.

Abdou had to start his new job the following week, and he had zero money for rent.

His mother gave him two pieces of her jewelry. His aunt gave him another piece. His father scraped together 100 Dirhams. Abdou and Marielle went to the city, sold the jewelry for 2,500 Dirhams, and took the bus to Rabat.

Abdou arrived at the ETAGOS headquarters at 8:00 a.m. He turned in his paperwork and joined ten other recruits in a conference room. Everyone else wore suits and ties. Abdou wore jeans, a polo shirt, and worn-out tennis shoes.

The HR manager pointed at him. "Ah, the American-educated one! The man with the field data experience. We are thrilled to have you."

"Thank you, sir."

"However," the manager added, eyeing Abdou's sneakers, "when you are at headquarters, we expect a suit and tie."

A man in a blue uniform served tea. He reminded Abdou of his father.

Just before lunch, the CEO walked in. "Which one is Abdou?"

Abdou raised his hand.

"Welcome to ETAGOS. I am Lahbil," the CEO said in English. "Come with me."

Abdou followed him into a splendid office filled with portraits of the King. Sitting there was Amir, an IAV alumnus from the class ahead of Abdou.

"Nice to see you, Abdou," Amir said.

"You will be based in Kenitra," the CEO explained. "In a few months, you will be in charge of the entire Northern zone. We have problems there I want you to solve. We will provide a car and a house."

"Thank you, sir."

Abdou began commuting from Rabat to Kenitra by taxi until Marielle returned to the US before Christmas. The Kenitra chief assigned him a driver.

Abdou's new office was full of file cabinets but no computer. He asked a technician for the climate, operations, and production data.

A day later, Amir gave him a tour of the massive farms. Over lunch, Amir leaned in. "Keep your eyes open. We are the only two from

IAV in this company."

Abdou didn't understand the warning yet.

He dug into the data. He verified the numbers four times because they didn't make sense.

First, the fuel costs for irrigation were impossible. More fuel was "spent" during rainy years than dry years. Second, the spending on herbicides and fertilizers was astronomical. Third, the records claimed they applied 500 kg/hectare of premium wheat seeds, when only 300 kg were needed.

Millions of Dirhams were disappearing.

Abdou called a technician to verify. The man looked terrified and insisted the data was correct.

When Abdou arrived at his office the next morning, he didn't recognize it. New carpet, new furniture, new empty file cabinets.

His graphs, notebooks, and data analysis were gone.

"Where are my papers?" Abdou growled at the security guard.

"A private company came to clean and equip the office, sir. They threw everything out."

Abdou was furious. He smelled a rat. He had asked the right questions to the wrong people.

That evening, as his driver dropped him off, the man opened the trunk. "Mr. Abdou, wait. I have two bags of clementines for you. Fifty kilos each. First harvest—not even in the market yet."

"No," Abdou said, his voice rising. "Don't do that again."

"But I can sell them for you! You keep the money."

"I am not corrupt! Who asked you to do this?"

"Everyone does it, sir. Everyone takes a share."

"I am not like them. I am not a thief."

"This is nothing. You should see what they do with the fuel and equipment."

Amir's warning suddenly made sense. The crooks were trying to implicate him. If he took the fruit, he was one of them. Abdou was grateful for Amir's advice and warnings.

Abdou called his former thesis adviser, desperate for a way out.

"Call Dr. Chafai," his adviser said. "There is a professorship opening at the ENAM agricultural engineering school in Meknes. The interview is tomorrow."

Abdou requested a leave of absence and rushed to Meknes with Marielle. They stayed with Mr. Boughanim, a kind friend from the village who worked as a bank auditor.

The next day, Abdou gave a presentation to thirty faculty members at ENAM. They grilled him for two hours. Ultimately, they unanimously approved his candidacy.

While waiting for the final paperwork, Abdou returned to ETA-GOS. He received his first paycheck.

At the same moment, ENAM called. He had the job.

Abdou took his ETAGOS paycheck, wrote "VOID" across it in big letters, and marched it back to the accounting department. He handed in his resignation and walked out.

He couldn't stand another minute among the thieves who had stolen his data and tried to buy his integrity with bags of oranges.

Abdou and Marielle moved to Meknes. They were broke, but they were free.

CHAPTER 17: The Professor and the Wedding

Abdou went to sign the paperwork for his new job as a professor at ENAM. He left Marielle at home and marched into the administration building, asking for the Dean's office.

The secretary blocked him. "The Dean does not meet with students."

"I am a new faculty member," Abdou explained.

She called the Dean, then hung up. "He still won't see you."

Puzzled and edgy, Abdou called his former adviser, Dr. Ouattar. Within minutes, a man in a gray suit appeared and ushered him in.

The Dean looked like an old rooster. He was a skinny, short man in his fifties who brushed his hair over a bald spot, giving it the appearance of a crest. The faculty nicknamed him "Houdhoud," after the Hoopoe bird.

"I want you to work with Mr. Brahim and Mr. Kandoor," Houdhoud announced without preamble.

Abdou nodded, though he didn't understand why until later. He submitted his paperwork and returned to the apartment.

It was freezing. Marielle, a Minnesota native, was shivering under layers of clothes. The apartment had no heat and no insulation; the temperature inside was the same as outside.

Christmas break arrived. Marielle flew back to Minneapolis, and Abdou visited his village. The house was full of people.

"How are ETAGOS and Kenitra?" a neighbor, who was a teacher, asked.

"I left ETAGOS. I am now teaching at ENAM."

"Are you crazy?" the teacher gasped. "You left ETAGOS? You had benefits! Grain, livestock, cars, fuel! And you traded it for chalk and a blackboard?"

"I am a university professor," Abdou replied sharply. "I teach engineers. I do research. I am not a thief. I left because the crooks there were doing exactly what you described—stealing."

"Yes, son," Abdou's father interjected quietly. "I am proud of you. We are not thieves."

The teacher clammed up.

Abdou returned to Meknes after New Year's, rejoining his generous roommate, Mr. Boughanim. Boughanim refused to take a penny for rent or food, knowing Abdou hadn't been paid yet.

Abdou threw himself into teaching. He taught four hours a day, covering everything from soil science to plant-water relationships. Boughanim, an economist, enjoyed reading Abdou's lecture notes.

One day, a colleague told Abdou, "We were in a meeting with the governor about drought mitigation. Boughanim stood up and said, 'Before we use the term drought, we have to define it...' He quoted you word for word."

Despite his hard work, Abdou was caught in a power struggle. A "clan" of Amazigh faculty members tried to recruit him. When he refused to join their faction, they reported him to Dean Houdhoud. In retaliation, Houdhoud froze Abdou's paperwork, preventing him from receiving his salary.

On March 3rd, the Feast of the Throne, Abdou was in the field taking measurements. His jeans were caked in mud, and his jacket was ripped at the elbows.

Two faculty friends ran up to him. "Your absence will be noticed, man. You are the only one not attending the festivities. The Ministry of Interior will go after you."

"I am all muddy."

"It doesn't matter. Come."

Abdou joined the line of faculty waiting to salute the Minister of Justice. Dean Houdhoud spotted him and rushed over, horrified.

"What is this? Don't you have a suit? Don't you have clean clothes?"

"Sir, I was in the field. You haven't signed my papers. I haven't been paid in three months. I want to celebrate the Throne, but I will go like this unless I get my paycheck immediately."

"Unbelievable!" Houdhoud hissed. "Come see me after His Excellency leaves."

An hour later, Abdou walked out of the Dean's office with a check for 11,000 Dirhams—a partial advance. He cashed it, bought dress shirts, pants, a sports coat, and black shoes.

He tried to pay Boughanim back for the rent and food. Boughanim got angry.

"Don't do that again. What I did is my duty. You are not in America," he said sternly.

Abdou apologized, grateful for his friend's traditional honor.

He showered, shaved, dressed in his new clothes, and returned to ENAM. He marched straight into Houdhoud's office.

"See? How do I look now? Did you think I didn't like to look like a professor?"

"Alright," Houdhoud grumbled. "I need to discuss your research."

"Not now. I have to go to the Ministry of Agriculture's office for the event. I just came to make a point."

Abdou realized that ineffective bureaucrats like Houdhoud viewed human capital as a burden rather than an asset. The Dean spent weeks alone in his office, speaking to no one.

Abdou's frustration mounted when Houdhoud denied his request for 75 kg of corn seeds, offering only 50 kg as if they were haggling in a souk. Abdou bought the remaining seeds and 32 aluminum tubes with his own money so his students could conduct their experiments.

He was broke again.

Marielle returned in June, complaining about the heat. "I am being cooked," she groaned.

In August, they went to Rabat to get married. They needed witnesses. Mo served as the translator, and Gall, an American professor of ill repute, served as the required Christian male witness.

But the judge refused to marry them.

"You need a Certificate of Virginity," the judge declared.

It was a misnamed document that simply meant Marielle was single. Abdou didn't need one because, under Islamic law, men could have multiple wives.

"There is no such document in America," Marielle explained.

They were sent from office to office, standing in endless lines. Finally, an old man approached them outside the court.

"My son lives in Canada. I heard your plight. I will help you. You don't need a judge. A Justice of the Peace can marry you. Meet me here at 4:00 p.m."

They met the man, who took them to a certified notary. Marielle wore a wedding dress her sister had made. She looked gorgeous.

The Justice of the Peace asked for her religion. "Catholic," she said.

He nodded. Islamic law allowed it. "How much dowry?"

Abdou and Marielle had agreed on 1,000 Dirhams. But knowing the official would take a percentage commission, Abdou quickly said, "500 Dirhams."

"But you said 1,000!" Marielle interrupted in French.

Abdou kicked her gently under the table and winked.

"What did she study?" the official asked.

"Sociology."

"Sociology *waloo*," the man muttered. *Sociology is nothing.*

He opened a book. "I will get a Christian witness. Alan approves for a 200 Dirham fee."

No one knew who "Alan" was, or if he even existed, but the marriage was certified.

"You are now really married, Madame," the official grinned. "Just close the curtains."

Abdou paid the 500 Dirham fee, 200 to the "witness," and 200 to the old man who helped them. Morocco ran on intermediaries.

Abdou and Marielle took the train to Marrakech to meet Mo and Christine. The train was packed. There were no seats, only bags of mail piled everywhere—likely heading to soldiers fighting the POLISARIO terrorist separatist front in the Moroccan Sahara.

Marielle sat on her suitcase in the aisle. Every time someone needed to pass, she had to stand and lift the suitcase over her head. Abdou slept on the floor.

They spent the night at the Hotel Mamounia, then took a terrifying bus ride over the High Atlas Mountains to Agadir.

Agadir looked brand new. After the devastating 1960 earthquake leveled the city, it had been rebuilt by an international coalition—French, American, and Moroccan forces. The architecture was a strange mix of beautiful styles.

They went to the beach to relax. Abdou noticed a circle of young Moroccan men laughing and pointing. He went to check.

Two topless French women were sunbathing, oblivious to the cultural shockwave they were causing. Abdou sighed and led Marielle away. They were definitely no longer in the village.

CHAPTER 18: The Decision to Immigrate

When Abdou came home from the university, he found a roasted chicken on the makeshift table.

"Where did you get that?" he asked Marielle.

"Long story! I went to the market with Said's young nephew. I bought a live chicken. The farmer killed it for us, but he didn't pluck it."

Apparently, Said's fourteen-year-old nephew had handed the warm, feathery bird to Marielle, expecting her to know what to do. Marielle had never plucked a chicken in her life. She shoved it back. The boy shoved it back. Since they didn't speak the same language, they played "chicken hot potato" until the boy finally sighed, plucked it, gutted it, and handed it back to her.

By mid-August, Abdou and Marielle had moved into their own apartment in Meknes. It was empty. They bought a mattress, a blanket, a thin carpet, a teapot, and a frying pan. Their dining table was a piece of plywood balanced on a box.

A week later, a friend who owned an appliance store sold them a stove and a used fridge on credit. No contract, just a handshake. "Pay me when you can," the Berber owner said.

Life in the apartment was noisy. They didn't own a TV, but their neighbors did. It was the summer of the 1984 Los Angeles Olympics, and Moroccans stayed up until 3:00 a.m. cheering for Nawal El Moutawakel and Said Aouita as they won gold on the track.

As soon as the TVs went off, the mosque loudspeakers crackled to life for the morning prayer. By 7:00 a.m., street vendors were shouting, *"Javeel! A Javeel!"* as they tried to sell homemade bleach.

Exhausted by the noise and the heat, Marielle returned to America in the first week of September to resume school at the University.

Abdou resumed teaching at ENAM. He befriended the librarian, who set aside new journals for him. He advised four students on their final thesis projects.

Then, his mother showed up unannounced. She brought his younger sister, Khadija, and his six-year-old brother, Majid.

"She needs to prepare for her Baccalaureate exams," his mother announced. "She will live with you."

At twenty-six, Abdou was suddenly raising a seventeen-year-old.

Majid looked around the apartment. "It's big, but where is the furniture? You don't even have a table."

"It's upstairs," his mother lied quickly, afraid word would get out that her engineer son was broke. "He saves it for guests."

She didn't know Abdou was only receiving half his salary due to the dean's bureaucratic games.

Abdou bought a small plastic ball and played soccer with Majid in the empty living room.

"See, little brother? This room is for soccer," Abdou joked.

"I wish we had an empty room for soccer in the village," Majid said.

Abdou bought a bed and supplies for Khadija. Her high school was within walking distance, but Abdou worried. Men stalked women relentlessly in Morocco.

One evening, coming home in the dark, Abdou saw two men following Khadija. She crossed the street; they followed.

He quickened his pace. As Khadija stopped in front of a store, one man leered. "Okay, beautiful Berber girl, *man-choufoukch*?" ("Can't we see you?") The other man pulled at her scarf.

Abdou snapped. Using his martial arts training, he laid both men flat on the sidewalk. A crowd gathered. One of the assailants was bleeding from the fall.

Suddenly, a *Mokaddem*—an Interior Ministry minder—pushed through the crowd. He took the attackers' IDs.

"Professor, go to your apartment," the minder told Abdou respectfully. "I know them. I will talk to their parents. May God protect you. Please go home."

Abdou was unsettled but not surprised that the minder knew he was a professor. The *Mokaddem* system knew everything—even how many loaves of bread a household bought.

Word got out. No one harassed Khadija again. But the neighbors weren't done.

A week later, Abdou came home to find Khadija bleeding, her face scratched. A neighbor and her maid had attacked her.

Abdou pounded on the neighbor's door. The husband opened it and shouted, "Stop bringing prostitutes to this building! This is for families, not a bordello!"

"What prostitutes?"

"First the English speaker, now this young Berber!"

"You son of a bitch," Abdou roared. "The American is my wife. The Berber is my sister. I will beat the shit out of you and your wife if you don't mind your business."

The neighbor slammed the door and locked it. Abdou waited, furious, but the man never came out. The next day, when the neighbor saw Abdou on the street, he ran away.

Abdou finally received his full pay, cleared his debts, and bought some real furniture. But the professional frustrations continued. Dean Houdhoud denied his request for 25 kg of corn seeds for research.

Abdou vented to Marielle in letters. They decided it was time for Plan B: Abdou would immigrate to the US.

But first, he needed a new ID and a visa. He needed to navigate the Moroccan bureaucracy one last time.

He had lost his National ID. To renew it, he needed a certificate of residence. The local *Moḳaddem* demanded a 100 Dirham bribe. Abdou refused.

Stalemate.

Abdou went to his landlord, Mr. Slimane, who lived across from the police station. Slimane walked him into the chief's office, and minutes later, Abdou had his certificate.

Next came the General Directorate of National Security (DGSN). While waiting in line, a plainclothes cop approached him.

"Hello, Professor. What does your mind desire?"

Abdou realized the depth of the surveillance. They knew who he was. A friend in training to be a *Caid* (sheriff) had once drunkenly confessed to writing reports on Abdou. "They know everything," the friend had slurred. "What newspapers do you read, the names of your friends..."

Abdou cut ties with that friend immediately.

A month later, his new ID arrived. Then came the packet from the US Consulate: "The US government gave you the status of family member to a US citizen.'

But it wasn't over. Abdou needed medical exams, X-rays, police records, and endless forms asking absurd questions: *Are you a communist? Do you believe in polygamy? Did you go on a honeymoon? Where? List of all organizations you were a member of?*

He gathered everything and went to the consulate in Casablanca.

Rejection.

"Your wife's income is insufficient," the clerk said. "Visa denied."

Abdou was devastated. He called Marielle. She was crushed. Her parents were suspicious—was he just using her for a visa? But eventually, her father sent a thick, detailed affidavit of financial support.

Abdou returned to the consulate. This time, he passed. They stamped his passport with a Resident Alien visa.

It was real. He was leaving.

Abdou sold his sofa and stereo. His brother took the rest of the furniture. Then, Abdou did something painful: he burned all his non-technical books, newspapers, and tapes. He couldn't risk being stopped at the border with "subversive" literature such as George Orwell's 1984 and Plato's Apology, and the Republic.

He went to the village to say goodbye. He didn't tell his parents he was leaving for good, but his brother, Mohamed, also an engineer, guessed the truth.

"You are leaving," Mohamed cried that night. "I will have to support the family alone."

"No," Abdou promised. "I will help as soon as I find a job. Don't worry, brother."

Back in Meknes, Abdou got a "vacation exit authorization" from the university HR department—a requirement for government employees leaving the country.

On August 18th, he received his final paycheck and went to buy a plane ticket.

"You need a round-trip ticket," the travel agent insisted.

"I am emigrating," Abdou argued, showing his visa. Finally, the agent sold him a one-way ticket to New York, with a connection to Minneapolis.

He called Marielle. "Arriving to MN from JFK on August 20th, 1985. 8:00 p.m. Northwest Airlines."

At 1:00 a.m., two of his students drove him to the airport in a car borrowed from a high-ranking military officer.

At border control, the police asked for his exit permit. They frowned at his one-way ticket but let him pass.

Abdou boarded Flight AT205 to New York. As the plane taxied, tears welled in his eyes. He was leaving his parents, his siblings, his friends, and the country he loved, heading into the terrifying unknown with nothing but the knowledge in his head.

He tried to sleep, but fear gave him stomach cramps. He wiped his eyes.

An elderly American couple in the next seat leaned over.

"Are you okay, young man?" the woman asked gently.

"I don't know. I just left my family. I am immigrating today."

She patted his arm. "Everything will be okay, my dear. Do you mind if we say a prayer for you?"

Abdou nodded, grateful for any blessing he could get.

CHAPTER 19: The New Immigrant

The two Americans sitting next to Abdou on Flight AT205 tried to engage him in small talk.

"It was very hot even in the early morning today," the woman, Kathy, said.

"Even the A/C in our hotel was having problems keeping up last night," added her husband, John.

"Were you on vacation in Morocco?" Abdou asked.

"No. We just stayed for one night. We came from Nairobi," John replied.

"How did you like Kenya?"

"Too much red tape and corruption. We were volunteers with Catholic Relief Services."

"My wife is Catholic," Abdou said with a slight smile.

"Is she in America?"

"Yes. In Minnesota."

"Are you a student?"

"No. I am an engineer. I was a professor in Morocco until yesterday."

"You are too young to be an engineer or a professor," Kathy commented.

"I am twenty-six. I studied at the University of Minnesota and in Morocco."

"Why did you leave? I would think Morocco needs people like you."

"The conditions were tough," Abdou sighed. "I like research, but the administration fought me tooth and nail over twenty-five kilos of corn seed. That, and many other reasons, pushed me to leave a country I love."

"John is an agronomist as well," Kathy said.

"Oh! What a coincidence!"

"What kind of work were you doing, John?"

"Helping small farmers improve yields with better corn seeds. Kathy was my data analyst. She helped using her Texas Instrument TI-54," John said, touching her knee affectionately.

"Hmm! I used a TI-59 with a statistical module."

When the plane landed in New York, Abdou said goodbye to the couple. He handed his passport and the large consulate envelope to the border control agent.

She opened the envelope, took his fingerprints and photo, smiled, and said, "Welcome to America."

Abdou retrieved his luggage, cleared customs, and found the shuttle to the Northwest Airlines terminal. He was exhausted and confused. He bought a bottle of water, leaving him with only $1.54 in his pocket. He still had 50 French Francs—money he had exchanged on the black market in Meknes because Moroccan Dirhams were worthless overseas.

He fell asleep on the flight to Minneapolis, waking only when the passengers applauded the pilot's perfect landing.

As he walked out of the gate, he saw Marielle. They hugged and kissed, clinging to each other after months apart.

They spent the night at a cheap hotel near the airport. The next day, they went to her parents' house. Abdou felt shy as Marielle reintroduced him to her family.

She took him to the Social Security Administration office. Within days, his Green Card and Social Security card arrived in the mail. He was amazed at the system's efficiency.

They stayed in a guest room in her parents' basement, near her father Bill's office. Abdou marveled at the shelves of books and scientific journals. Marielle's mother, Carole, took his 50 Francs to the bank and exchanged them for dollars.

Abdou had brought gifts: a Moroccan *foukia* (a long male robe) for Bill and cheap kaftans for the women.

That weekend, Marielle's sister Anabella hosted a barbecue. Abdou was touched to see everyone wearing the Moroccan outfits to welcome him. Even Bill wore his *foukia*, though it was comically short for his tall frame.

The following weekend, Bill hosted a brunch. Abdou was surprised to see the family drinking champagne at 11:00 a.m. *A real welcome party,* he thought.

During the week, Abdou was alone in the house while Marielle worked at a hotel. He spent hours hand-writing his resume or reading Bill's magazines. Whenever he heard footsteps, he ran downstairs to his room, too shy to interact until he grew more comfortable.

Marielle soon received a job offer from a market research firm, finally putting her sociology degree to use. But she kept her hotel job because Abdou hadn't found work yet.

By mid-September, they moved into their own apartment on Lake Street and Bryant Avenue South. It was a tiny unit near a police station in a run-down area. Bullet holes marred some of the neighboring windows. They were lucky to live on the second floor, but police searchlights swept through their bedroom almost every night.

Walking Lake Street in the 1980s was eerie. Buildings were boarded up and dilapidated. Homeless people were everywhere. Abdou couldn't understand it. Back in his impoverished village, even the poorest had a safety net of family and neighbors. Here, people were left to rot on the sidewalk.

His fear grew with every homeless person he saw. *This could be my fate,* he thought. *What if Marielle loses her job?*

He watched an old couple pushing a stolen "Rainbow Foods" grocery cart filled with their belongings. His heart raced. He felt he had arrived too late; there was no future for him here.

Their upstairs apartment caretaker didn't help. She stomped on the floor day and night, even when Abdou and Marielle were silent. Abdou nicknamed her "The Runaway Cow."

Abdou went to the Hennepin County Library, copying company addresses from phone books. He sent hundreds of resumes. Silence. Occasionally, a rejection letter. Although he had a US education, he had no US work experience and no references.

Desperation set in. Depression followed.

Marielle worked in her research job from 8:00 to 5:00, then in her hotel maid job from 5:30 to 9:30. She got home at 10:00 p.m., exhausted. They ate dinner late, and she was up again at 6:00 a.m.

Then came the letter from his brother in Morocco.

"Dear brother... The family is worried. You promised to send money. But you don't even send letters. You left the entire responsibility to me. I don't make much, but I contribute what little I have... Signed: Your miserable brother."

Abdou read it and burst into sobs. The guilt was crushing. He couldn't support his parents financially. He couldn't even help himself. His wife was working two jobs while he sat at home. He felt sad that he left his brother doing all the heavy lifting.

On a brutally cold November morning, Abdou walked to McDonald's. He applied for a job flipping burgers.

The manager sat him down. She looked at his resume—the engineering degree, the professorship, the research.

"Sir, I am sorry," she said gently. "This place is not for a person with your education and skills. You have no future here."

"But I need a job to eat."

She looked at him with pity. She bought him a Big Mac.

Abdou ate the burger, grateful for the food but devastated by the charity. He didn't need pity; he needed work. His self-worth was shattered.

He went home and cried. Marielle held him. "Do you want to go back to Morocco?" she asked softly.

He thought about it. But he couldn't go back. He had quit his government job for life; he could never get it back. He was stuck.

He watched the news. The 1980s Farm Crisis was destroying rural America. Willie Nelson was organizing Farm Aid concerts. Every day, TV stations showed farm auctions—families losing everything. The bankruptcies touched Abdou to his core. He was an agronomist watching American agriculture collapse.

He felt he was in the wrong place at the wrong time.

But he remembered the words of his mother and grandmother: *The brave horse rider always gets up, dusts himself off, and fights again.*

So Abdou got up. He dusted himself off. And despite the unbearable pain of rejection, he tried again.

CHAPTER 20: The Porter's Price

Around mid-November, Abdou was walking along Lake Street when he saw a sign: **Norell Temporary Employment Services.**

He went inside and filled out an application. A few days later, the phone rang at 4:00 p.m. The woman on the line told him to show up at 4:30 a.m. the next morning. He would share a ride to the job site, but he had to pay a $2 transportation fee. The pay was $3.35 per hour.

Abdou was desperate. Marielle hugged him and dropped him off at the Norell office in the pitch black of early morning. He was half-asleep and drowsy.

He introduced himself to the receptionist. She pointed him toward two other men. The three of them piled into a rusty old car and drove out of Minneapolis into the darkness of the countryside.

Abdou sat in the back, silent and afraid.

The driver lit a joint. The smell of marijuana filled the car. He took a drag and passed it to the man in the passenger seat. After a few puffs, the man turned and offered it to Abdou.

"No, thank you," Abdou said, trying to sound tougher than he felt. "I just had a few for breakfast."

He hadn't had breakfast. He hadn't smoked anything. He was just terrified.

They arrived at a warehouse as the sun began to rise. A supervisor was waiting. He led them to a long semi-truck parked at the loading dock.

"You are going to unload this baby," the supervisor said. "Put everything inside the warehouse."

They started unloading. The truck was filled with bags of insulation material. They were bulky but light.

Abdou grabbed his first bag. As he walked it into the warehouse, tears streamed down his cheeks. He avoided eye contact with his coworkers.

From university professor to porter, he thought. *This is who I am now.*

In Morocco, this job was called *Talb maachou*—"he who seeks his livelihood." It was the lowest rung of society, the work of the desperate.

He kept working. The other two men, high from the drive, found a stack of insulation bags in the back of the truck and fell asleep. Abdou worked alone.

By 10:00 a.m., the supervisor returned. He saw the two men sleeping and Abdou sweating, moving bags by himself.

He handed Abdou a cold can of Coca-Cola.

"I am sorry you had to do this by yourself," the supervisor said.

"Where are you from?"

"I am from Morocco."

"Hmm! I was in Marrakech in the 1970s, during the Vietnam War. Nice place." He patted Abdou on the shoulder. "I have to go back inside. You finish up."

Abdou finished unloading the entire truck by lunchtime.

On the drive back to the employment agency, Abdou's mind drifted to his father. He remembered stories of his dad wandering from farm to farm on a rusty red bicycle, begging for work after his French employer fled Morocco following independence. The cycle of poverty felt inescapable.

The car stopped. Abdou didn't realize they had parked until the driver turned around.

"Hey, man, are you okay? I saw you crying on the job site today. And on the way here. Take it easy."

Abdou thanked him and stepped out onto Lake Street. He walked in a daze, trying to find his apartment building among the rundown storefronts.

When he got home, there wasn't much to eat. He made a piece of toast and a glass of tea—the same meager lunch he used to eat in the village of Timoulilt. His life had rewound twenty years.

He sat on the couch, head in his hands. *Did I make a mistake coming here?* The more he thought, the more wounded and confused he felt.

He called his friend Moha in Chicago. Moha was an elder from his village, a former English teacher in Beni Mellal who had immigrated in 1975. Rumors in the village claimed Moha was CIA, but he was just a man trying to survive, like Abdou.

Abdou sobbed into the phone.

"Things will eventually get resolved," Moha consoled him. "All immigrants go through what you are going through. Go for a walk. It will help."

Abdou hung up. He remembered his grandmother's words: *"Warriors fall and get up."*

He wiped his face and went for a walk. An hour later, he saw a "Help Wanted" sign in a window: **Chemistry Technician**.

He walked in and asked for an application.

Two weeks later, Abdou received his paycheck from Norell Temporary Services.

He tore open the envelope. After taxes, Social Security, and the $2 transportation fee, the check was for exactly $12.20.

He took the check to the grocery store. He bought two tomatoes, a loaf of bread, a bag of lentils, and one turkey drumstick.

As he walked home, clutching his bag of lentils, he remembered the words of the border control agent in New York.

Welcome to America.

CHAPTER 21: The Law Has Changed

Abdou filled out the application for the chemistry technician position and handed it to the receptionist. During the interview, a young man wearing plastic boots and blue overalls explained that his job involved mixing chemicals to make shampoo.

He asked for references. Abdou had none, except for his graduate school adviser. Abdou noticed the young man looked as malnourished as he did.

He never heard back.

A month before Thanksgiving, his sister-in-law, Annabella, called. "Do you want to work at Montgomery Ward? I can get you an interview."

Abdou put on his best suit and took the bus to Southtown. Annabella's boss interviewed him and offered him a job in the men's department.

At first, the hours were good. But after Christmas, business dried up. His schedule was cut to just a few hours on Saturdays and Sundays. The pay hardly covered his bus fare. Every day for lunch, he ate a peanut butter sandwich.

One afternoon, the phone rang. Abdou's heart leaped—he thought it was a job offer.

"Hello, how are you doing, sir? Enjoying the day? This is Mark."

"Hello, Mr. Mark."

"I am calling to schedule a visit with you. Are you available tomorrow?"

"Yes, Mr. Mark."

"What is your address?"

Abdou gave him the address and set the appointment for 8:00 p.m. He hung up, excited. He told Marielle that a man named Mark was coming.

"Who is Mark? Why is he coming?" she asked.

"He is coming because the law has changed."

"What law?"

"I don't know. But it sounded important."

Mark arrived the next evening wearing a suit and carrying a briefcase and a thick binder. Abdou, Marielle, and Mark sat around the small kitchen table.

Mark opened his binder and started flipping chart after chart, table after table. He went on like a broken record. It took Abdou ten minutes to realize that Mark wasn't offering a job or legal aid—he was trying to sell them life insurance.

"Mr. Mark," Abdou interrupted. "My life is worthless. I have no money and no job."

"But you said you are an engineer?"

"A broke, unemployed, and broken soul, sir."

Mark didn't stop. He kept flipping charts. Abdou realized there was no escape, opened a statistics textbook, and started reading silently while Mark droned on.

Finally, Marielle stood up. "Please leave," she told Mark.

She was furious with Abdou for inviting the salesman in. For years afterward, whenever Abdou got his hopes up, she would joke: "The law has changed."

Around March, they moved to a cleaner basement apartment owned by a high school teacher.

Abdou kept working his weekend shifts at Montgomery Ward. He befriended an older man who came in every weekend to buy the same brand of short-sleeve shirt in a different color. It was the only sale he made consistently.

He secured another temporary job at Pillsbury Food Company as a technician. It took two bus transfers to get there. His job was to weigh ingredients and record data for a new pizza product.

One day, he walked into a giant industrial freezer to retrieve ingredients. The heavy door clicked shut behind him.

Abdou pushed the handle. Locked.

Panic set in. The cold bit through his clothes instantly. He pounded on the heavy steel door, screaming for help, but the insulation swallowed the sound. He was trapped for thirty minutes, his body temperature dropping, before a coworker finally opened the door.

Abdou stumbled out, shivering and terrified.

The next day, his position was terminated. The company called it a "near-miss safety event." Abdou called it getting fired for almost dying.

Abdou and Marielle saved every penny, barely keeping their heads above water. Abdou was sinking deeper into depression, though Marielle remained his rock.

He received a letter from his father saying the entire family missed him. Abdou read it and imagined the sadness on his mother's face.

He wrote back less and less. He was too embarrassed to tell them the truth about his life in America. On rare occasions, he put a $20 bill in an envelope addressed to his father. In the short notes, he lied and said he was busy studying. He couldn't bear to let them know that their engineer son was a porter, a clerk, and a man who got locked in freezers for minimum wage.

CHAPTER 22: The Pivot to Land a Job

Abdou stopped applying for labor jobs and pivoted. He spent his days reading about statistics and studying a FORTRAN programming book his father-in-law gave him. Bill also gave him a challenging document on vector processing. Abdou devoured them, then checked out *Programming with Pascal* and *Computer Simulations and Modeling* from the library.

On his day off, he went to the University of Minnesota bookstore and bought a thick tome called *Programming with Turbo C*. He studied it every day, realizing that his path back to a career lay in computers.

He had about $400 in savings. One morning, he saw a newspaper ad for an Atari 130XE computer and a printer for $350. He ordered it by phone, Cash on Delivery (COD). He didn't tell Marielle he had used their emergency funds, but he knew she would forgive him if it led to a job.

A week later, two boxes appeared in front of his apartment door.

He opened them. One contained the computer, the other a small thermal printer. He looked for the invoice to pay the COD.

There was no bill.

He called the company. A recorded voice answered: *"The number you are calling has been disconnected. No further information is available."*

Oh crap, Abdou thought. *They went out of business.* They had shipped the inventory but never collected the cash.

It was the first stroke of luck he'd had in a year.

Abdou spent his days at the library or at home, writing code for statistical packages on his free Atari. In the spring, he rode his bike around Lake Calhoun and Lake of the Isles, clearing his head before returning to his self-imposed boot camp.

The long year of unemployment had taken a toll. The American Dream had turned into a nightmare of rejection and poverty. Abdou applied for everything from lab technician to scientist, willing to swallow his pride just to put food on the table.

Then, the phone rang.

It was a secretary from the Department of Ecology, Evolution, and Behavior at the University of Minnesota.

"Dr. David Tilman would like to interview you tomorrow," she said.

Abdou froze. Dr. Tilman was a world-renowned scientist heading the Long-Term Ecological Research (LTER) project, funded by the National Science Foundation (NSF). Abdou didn't even remember applying for the position, but he wasn't going to ask questions.

He put on his suit and took the bus to the university. He was nervous as he entered the department building next to Coffman Union.

The interview began. Dr. Tilman sat with Robert, the man currently holding the position.

"The person we hire will manage all the data for LTER," Dr. Tilman explained. "They will manage our technology and experiments, and support our modeling and system simulation efforts. We have an experimental station about fifty miles from here."

"You indicated that you have a minor in statistics," Robert said. "Tell us more."

"My transcript shows I took graduate courses here at the U of M," Abdou replied. "I also took four years of statistics and biometry in Morocco."

"Do you know FORTRAN?" Dr. Tilman asked.

"Yes. I used it as an undergraduate and for writing simulations in graduate school. I've also been studying FORTRAN in vector processing on Cray supercomputers."

"Tell me about your experience analyzing data," Robert pressed.

"I use basic statistics, ANOVA, linear and multiple regression, and non-linear regression. I also write my own statistical software packages."

"I saw in your resume that you did a lot of work in experimental design," Dr. Tilman noted. "Tell me about your soil science experience."

"I designed many experiments at the University of Nebraska and at ENAM in Morocco," Abdou said, listing the designs—randomized blocks, split plots. "My soil science experience ranges from physics and mechanics to analyzing total nitrogen, organic matter, and ammonium nitrate."

The interview lasted two hours. They pushed him hard on the details of statistics and soil science.

Finally, Dr. Tilman stood up. "Well, it was nice talking to you. I have to be honest—we offered the position to a woman. She is hesitating. If she does not accept it in two days, we will call you."

"Thank you, Dr. Tilman. Nice to meet you, Robert."

Abdou left, his heart pounding. He learned that Robert was going to get his Ph.D. in statistics at the University of Illinois. Abdou desperately wanted to replace him, but he worried his answers hadn't been enough.

Abdou went home and began the vigil. He waited by the phone day and night.

Each time Marielle called to check in, he panicked. "Don't call!" he begged. "Leave the line open in case they call!"

Two days passed. Then three. No call.

On the fourth day, the phone rang.

"Hello, Abdou. This is Dr. Tilman."

Abdou held his breath.

"The position is yours if you want it. Your job title is Junior Scientist and Data Manager. The starting salary is $18,000 per year. You start next week, July 15th."

Abdou closed his eyes. He was desperate. He had no money. But he remembered his worth.

"Yes, I would love to accept the position," Abdou said. "But I would like to get paid $21,000."

A brief silence followed. Abdou's heart hammered against his ribs.

"The best offer for this position is $20,018," Dr. Tilman said, consulting a pay schedule. "That is the limit."

"Thank you. I accept."

"Come by the department today or tomorrow. Ask Arlene for the paperwork."

"Yes, sir. Thank you."

Abdou hung up and let out a shout. He rushed to the university immediately. Arlene made a copy of his green card, and he filled out the forms with a shaking hand.

He called Marielle. They celebrated that night like they had won the lottery. The relief was physical. Marielle quit her hotel job immediately to focus on her career in market research.

Abdou started his new job at the University of Minnesota on Tuesday, July 15th, 1986.

As soon as he received his first paycheck, he bought an old Chevy Monza. He needed it to drive to the experimental station. He was no longer a porter. He was a scientist again.

CHAPTER 23: Ali is Here

Abdou visited the LTER experimental station with Robert, who gave him a tour of the lab and the field experiments. Abdou was impressed by the sheer scale of studies on biodiversity, interspecies competition, the effects of soil nitrogen on biomass, and fire frequency.

However, Abdou's agronomist eye noticed something missing. There were no experiments on water stress or competition for water, despite the obvious signs of drought on some plots.

Robert spent a week with Abdou on the Minneapolis campus, dumping a massive amount of knowledge about datasets and processing applications before leaving for his PhD program in Illinois.

Abdou was thrown into the deep end. He met Cliff, the soil chemistry manager, and Andrea, the graphics designer. He met dozens of graduate students and postdocs. There were too many names and faces to remember.

A couple of months later, the lab manager quit. His replacement was arrogant and sloppy, clashing with everyone. He lasted a month.

Dr. Tilman called Abdou into his office. "Can you run the lab as well?"

Abdou accepted. He hired a technician and undergraduate students to help. His job was reclassified to Associate Scientist—a promotion that came with a raise, though he still wasn't back to the level he had enjoyed in Morocco.

On Christmas evening, the University police called Abdou at home. A pipe had frozen on the fourth floor. Water was everywhere.

Abdou spent his holiday night moving computers, file cabinets, and thousands of floppy disks out of the flood zone. The losses were staggering.

He spent months restoring data. He and Tilman wrote a grant to replace the hardware. They upgraded from IBM PCs to Sun Microsystems servers, bought SAS for statistics, and installed a C++ compiler.

Abdou built a network for the LTER team using "thin-net" coaxial cables. It was finicky; if a student kicked a cable under a desk, the entire network went down. Abdou had to learn networking on the fly, relying on the university's IT staff, and his brother-in-law Mike for help.

Hungry for more skills, Abdou applied to the Computer Science program. He was rejected. The admissions officer claimed his transcripts lacked math and physics—even though Abdou had taken more advanced math in graduate school than most applicants. Moroccan transcripts were sloppy and did not give details about his coursework.

Devastated but determined, Abdou tested out of several courses, earning more credits than required. He applied again. This time, the Institute of Technology admitted him.

Staff tuition was free, so Abdou took classes at night and during his lunch breaks.

At work, he bonded with David, a PhD student from Iowa. They spent their breaks listening to the Iran-Contra hearings and dissecting Oliver North's testimony.

"I built an igloo this weekend," David mentioned one day.

"What is an igloo?"

"A snow hut. I slept in it."

"You dudes prefer playing in the snow to making babies," Abdou joked.

"I know you have snow in Morocco," David said.

"Yes. We have snow, deserts, mountains, and two oceans. It is like California."

David once shared *lutefisk* with him. It was awful—fish treated with lye until it became gelatinous—but Abdou appreciated the gesture.

"Are you worried about the NSF funding cuts?" David asked later.

"Yes. Especially since we are expecting a baby."

Things looked gloomy when the Reagan administration cut funding for science. Abdou was on "soft money"—his paycheck depended entirely on grants. Fortunately, the university stepped in to bridge the gap.

Abdou wrote code for simulations on a Cray-2 supercomputer. He also rewrote a complex ecological model developed by Princeton professor Stephen Pacala, optimizing it to run on their hardware.

Despite his success, Abdou kept telling Tilman that the field experiments showed severe water stress.

"You are just thinking about Morocco," Tilman joked.

But Abdou was right. Minnesota experienced a severe drought in the spring and summer of 1988. The impact on biodiversity was undeniable. Abdou registered for an independent study with Tilman and built a long-term climate dataset dating back to the 1860s.

When Tilman saw the report, he was excited. They submitted a paper that became one of the first to link global climate change to species loss.

Abdou was promoted to Scientist and Data Manager.

Then, in the summer of 1988, Abdou received a call at 5:00 p.m.

"Hello, cousin, how are you? I am here," a familiar voice said.

"Ali? Where are you?"

"I am in New York."

"Are you going to Utah again?"

"No, I am coming to visit you. My plane is boarding. See you in three hours." Click.

Abdou stared at the phone. He didn't know the flight number. He hadn't been warned. But this was the Moroccan way. People simply showed up without announcement.

Marielle, seven months pregnant, was less charmed. They ran

around like headless chickens, cleaning the guest room and buying food.

Abdou drove to the airport. He found Ali standing near the luggage carousel wearing a heavy wool suit and a winter jacket.

Abdou burst out laughing. "You look like a frozen duck! It's July! Take that jacket off."

"But you told me Minnesota is cold," Ali said suspiciously.

"Not in the summer!"

Ali refused to believe him until he stepped outside. The humidity hit him like a wall. The heat index was 105.. Ali gasped for air and finally stripped down to his soaked undershirt.

Ali spent his days walking around the Linden Hills neighborhood in his dress clothes while Abdou and Marielle worked.

On Saturday night, Abdou took Ali to a bar on Lake Street. Within minutes, Ali disappeared. Abdou found him at the counter, laughing with a woman named Amy who was taking his picture.

When Abdou approached, Ali switched to Tamazight. "Leave us alone, cousin."

"Is everything okay?" Abdou asked the woman.

"I'll take him home," Amy said. "Give me your address. I like him already. I will bring him to your home."

Ali returned the next day at noon.

"So?" Abdou asked.

"We went to her house. A mansion on the lake! She drives a Mercedes. A real *djaja b kamounha* (a chicken ready to eat)."

"And?"

"She handed me a toothbrush. I told her I have one at home. I put it in my pocket. She laughed hysterically. She grabbed it, put toothpaste on it, and handed it back. She laughed even harder. Finally, I realized she wanted me to brush my teeth right there."

"*Rask Tkeel*," Abdou laughed. "You are slow, cousin."

Ali wanted to see her again, but he hadn't asked for her number or her last name. He spent days scanning the phone book for "Amy" and riding Abdou's bicycle to the bar in his suit, but he never found her.

One day, Marielle turned on the oven to preheat it for a pizza. Smoke filled the kitchen. The stove caught fire.

Ali had cooked lamb chops directly on the oven racks without a pan. The fat had dripped everywhere.

Exhausted, Abdou took Ali for a walk around Lake Harriet. Ali pulled two garbage bags from his pocket.

"What is this for?"

"You catch a duck. I will catch a big goose."

"No, Ali!" Abdou said firmly. "We are not catching geese. This is not Morocco. America has laws against what you want to do poacher."

"Are you soft now, cousin?"

"No. I am law-abiding. Let's go home."

Abdou and Marielle went to the experimental station at Lake Itasca for a week, leaving Ali alone. When they returned, they found every window open and the air conditioner running full blast. The electric bill was over $450.

A week after Ali finally left, the phone bill arrived: $500. Ali had been calling 1-976 phone sex numbers and making long-distance calls to friends in Morocco at $2.65 a minute.

Ali's visit was an expensive whirlwind.

In January 1989, Abdou was sworn in as a US citizen. He held his newborn son at the ceremony in the federal court, led by St. Paul Mayor George Latimer.

He had passed his Computer Science GREs with flying colors. He applied to graduate school again. He was ready for the next chapter.

CHAPTER 24: The Government Grind

Abdou waited months to hear about his graduate school application. Finally, he requested a meeting with the Director of Graduate Studies in Computer Science.

To his surprise, the director was the same racist grumpy professor who had taught his Systems Organization and Assembly Language course. The old man glared at him as he entered.

"What do you want?" he groaned.

"Professor, I am here to check on my application."

"I have nothing to tell you. Be patient. I don't have time for these meetings. Don't bother me again. I don't know why they keep scheduling these meetings."

Abdou left, fuming. *What a jackass.*

In the hallway, he ran into Erach, an AI PhD student who worked for him.

"Hello, Erach. How goes it?"

"Horrible. Did you hear about the flyers? Someone put racist slurs in the mailboxes of brown faculty and students."

"Who? Why?"

"It's bad. Professors Sahni, Ibarra, and Powell all left."

Abdou was shocked. Those were top-tier professors. He decided to freeze his application but continued taking courses as a non-degree student.

Meanwhile, he was promoted to Senior Scientist at the university. He and Marielle moved to a new house, close to both campuses. Marielle took a job as a researcher at the Wilder Foundation, tracking data on homelessness, the elderly, sexual abuse, and program evaluation.

Eventually, Abdou reactivated his grad school application. He met with the new Director of Graduate Studies, Dr. Maria Gini.

"I see you have a graduate degree in soil physics," Dr. Gini said. "My PhD was in physics, too. And you're a Senior Scientist at the Minnesota Supercomputer Center?"

"Yes. I am in ecosystems. Working on system simulations."

"I am impressed. I will fast-track your application."

With her support, he was accepted. Abdou initially joined the networking group but switched to distributed computing because the networking students conducted all their research discussions in Chinese, which he didn't speak. He focused his thesis on Parallel Geographic Information Systems (P-GIS), and distributed computing.

However, Abdou's university pay stagnated. With another baby on the way, he needed stability. He applied for a position with the National Oceanic and Atmospheric Administration (NOAA) to work on airborne systems and remote sensing.

He got the job. His new office was in Minneapolis-St. Paul International Airport.

On his first day, Commander Bob took him to the nearby Air Force base for fingerprinting. Then came the meeting with the pilots. The room was filled with swearing. Abdou, still shy, blushed.

"Look at him!" one commander laughed. "He is blushing! We need to change that."

That evening, a Lieutenant Commander approached him. "Let's go. Important meeting off-site."

They drove to a local bar near a church. The entire pilot team was there.

"This is a special training session," Commander Max announced. "Repeat after me: Fuck, fuck, fuck!"

Abdou stared at his beer, mortified.

"You want to be part of the team or not?" Commander Bob asked. "Get used to the swearing. Culture matters. Tell the commander to go shit in his pants."

It was hazing, but friendly. Finally, Abdou raised his glass.

"Fuck you all! Are we a team now?"

The table erupted in cheers. "Welcome aboard, son of a gun!"

Abdou quickly proved his worth. He discovered the airborne systems were wide open to hackers. He performed a penetration test, modifying data from an external network. The director went through the roof and immediately assigned him to shore up security.

Soon, the office would relocate to a new facility in Chanhassen. Abdou asked about the network design.

"What network? Aren't we just running a wire?" the director asked.

"No, sir. We need a modern infrastructure."

Abdou designed a state-of-the-art network—raised floors, suspended cable trays, massive UPS backups. The bureaucrats in Kansas City balked at the cost, but the director pulled strings and got it approved.

Abdou then tackled the aging airborne system: three small planes equipped with nuclear detectors and ancient HP 85 computers. The software was written in HP BASIC, a tangled spaghetti code mess.

He had to rewrite it in C. Terrified of introducing bugs that could cost lives, Abdou wrote a Perl script to automatically parse and convert the code.

Testing was rigorous. Abdou built dry-ice boxes to simulate high-altitude temperatures and used radioactive lantern mantles to test the sensors. But the only way to test the GPS and altimeters was to fly.

Abdou flew with the pilots—Captain Barry, Commander Andrea, and Steve, a former Apache helicopter pilot who had flown in the Gulf War.

"Take me down to 8,000 feet," Abdou requested during one test flight.

"Here we go!" Steve yelled. The plane dropped 3,000 feet in seconds. Abdou clung to his seat, feeling his lungs try to leave his body.

"Are you okay?" Steve asked over the headset.

"I was afraid."

"Man up, Abdou! At least no one is shooting at us! Don't soil your pants!"

The final test required flying 164 feet above Lake Mille Lacs to calibrate the zero gamma radiation reading (deep water blocks gamma radiation). Abdou begged Barry and Rob to fly it. He trusted Barry, a hurricane chaser, not to crash into the water.

Local fishermen were not thrilled to see two planes buzzing their boats, even if the local authorities were notified about the flights over the lake.

The pilots, the director, and Abdou analyzed the data independently and everything matched perfectly. The system was approved for deployment.

But despite his success, the grind of government work wore him down. He discovered that a colleague he supervised—a man with less education—outranked him in pay grade.

One day, another redneck colleague sneered at him. "Hey, boy!"

Abdou snapped. He pulled the man outside. "You call me 'boy' again, and I will stick my fist up your ass."

"No, no! In the South, we call people we like 'boy'!"

"I know what you did in the South. Don't talk to me again."

Abdou started a consulting company on the side, Hawk Technology, working evenings and weekends, building and installing firewalls. He made more money part-time than he did at his full-time job.

The final straw came when a file server crashed due to a full disk. Abdou investigated and found a system administrator hosting pornography on the government server. He reported it.

Instead of firing the admin, the "Hey Boy" colleague got mad at Abdou for putting the report in an email—creating a paper trail.

"Abdou, this place does not deserve you," a great pilot friend told him. "Go to the private sector."

Abdou defended his Master's thesis in computer science and applied for a job at Bridge Information Systems, a Wall Street IT, and data firm.

They interviewed him, took him to dinner, and offered him the job.

Abdou accepted the position. It was time to leave the bureaucracy behind.

CHAPTER 25: The Bridge to Freedom

Abdou resigned from the National Remote Sensing Center with three weeks' notice. The three weeks were like hell for him.

On his first day at Bridge Information Systems, he arrived at 7:30 a.m. expecting a bustling office. Instead, he found only construction workers.

"Where are the engineers?" he asked.

"We don't know. Someone hired us to renovate," the foreman shrugged.

Oh crap, Abdou thought. *The company went bankrupt. I resigned from the government for nothing.*

Panic settled in. He stood there, sweating, imagining telling Marielle he was unemployed again.

Fifteen minutes later, engineers started trickling in. When Abdou told them his fears, they burst out laughing. They showed him a temporary open area where the team was working during the renovation. The incident became an instant office joke. His teammates nicknamed him "The Body," after Minnesota's wrestler-turned-governor, Jesse Ventura.

Abdou loved the new job. His boss, known by his handle "TL," and many team members were veterans of Cray Supercomputers. TL was a brilliant engineer who had helped port UNIX to the Cray operating system.

Life took a turn for the better. Abdou learned about investing. He was treated with respect. By Christmas, despite having been there less than a month, he received a generous bonus. He had more than doubled his government salary.

But the work was grueling. Abdou was a Senior Engineer for the Wall Street firm, and the markets never slept. For six months, his pager went off at 3:00 a.m. and 4:30 a.m. every single night.

Seeing his exhaustion, TL and the Executive VP, Chris, hired another engineer—a man with a PhD in Computer Science from a top university—to help.

Abdou onboarded him and handed over the pager. They agreed to alternate weeks on call.

On the PhD's first week, Abdou went to bed hoping for a whole night's sleep.

Beep-beep-beep.

The pager went off at 3:00 a.m. The Operations team couldn't reach the PhD. Abdou fixed the problem.

Beep-beep-beep.

4:30 a.m. Same thing.

This continued all week. Each morning, the Operations report read: *"We paged the PhD several times. No answer. We paged Abdou, and he resolved the issue."*

TL apologized repeatedly. They tried talking to the PhD.

Finally, the man snapped at Abdou. "Abdou, I have a PhD. I should not be paged."

"Your fucking job includes responding to pages," Abdou retorted, his patience gone. "I will not cover for you anymore."

TL and Chris didn't hesitate. Chris flew in from New York and fired the PhD.

A week before Christmas, Chris and TL handed Abdou two massive checks—a bonus and a 35% raise.

They celebrated the holidays at *La Belle Vie*, a fancy French restaurant. One engineer ordered a $4,000 bottle of wine. When Chris saw the bill, he whispered, 'Oh, shit," then calmly asked the manager to split it four ways to hide the high cost from corporate oversight. The waiters got a massive tip.

Thanks to TL, Abdou's financial success allowed him to double his remittances to Morocco. He brought his sister to the US to study, though she struggled with Computer Science because of the math requirements and eventually switched to Business.

Then, he decided to bring his parents for a visit.

He bought them tickets from Casablanca to Minneapolis via Paris. But Air France redirected the flight to Boston. Abdou had given them a note in English that read: *"Please help me. I am lost. Call my son."*

A kind stranger in Boston helped them through customs and to their gate to Minneapolis, then came a call from an airline stewardess.

"Please speak to your father. He is very agitated. He thinks he is lost."

"Hello, father. You aren't lost. You'll be here in three hours."

"This is no time for jokes, son!" his father shouted. "We flew all day, and the sun never set! We are lost!"

Abdou calmed them down. He met them at the Minneapolis airport. His father wore a blue suit and shined black shoes. His mother wore traditional Amazigh clothing and a headscarf, looking like a nun. She burst into tears and hugged him.

"Where are my grandchildren?"

"At home, Mom."

When Abdou pulled into the driveway of his house in Roseville, his parents looked confused. "Why are you parking here?"

"This is our house."

His father's face broke into a massive smile. "*Al hamdou Li Allah*," he whispered. *Praise be to God.*

He later admitted he assumed Abdou lived in a slum, knowing how hard life was for immigrants in France.

They entered the house, removed their shoes, and whispered *"Bismillah."* – in the name of God.

Cultural clashes began immediately.

One day, his parents sat him down for an intervention. "Son, your boy spends all his time in front of a machine that looks like a TV. He needs to go outside and play!"

"That is a computer, Mom. He is playing games."

They had never seen a computer. There was no word for it in Tamazight. They watched Abdou working from home, typing furiously.

"We know you are an engineer," his mother asked, "but why did you become a secretary?'

Abdou sighed. "Dad, give me a list of huge numbers."

His father wrote them down. Abdou typed them in and produced the sum instantly.

"It automates math," Abdou explained.

His father smiled. "*Laajab!*" *Amazing.*

His father fell in love with The Home Depot. He walked the lumber aisles, looking at the wood and muttering, "Ba ba ba ba."

"Why *ba ba ba*, Dad?"

"It is a gift from God. You can make anything from this wood."

But not all American customs went over well.

One afternoon, Abdou's friend, Steve, arrived with his girlfriend. They had been windsurfing. The girlfriend walked in wearing a tiny bikini, and partly exposed large breasts.

Abdou's sister intercepted them at the door, panicked. "*Yalatif! Yalatif!* She is *aariana* (naked)! Go upstairs before the parents see!"

Later, Abdou took his parents to the Mall of America. They walked past a group of teenagers dancing dirty to a boom box, midriffs exposed.

"*Yalatif, Yalatif,*" his mother chanted, covering her eyes.

Abdou tried to buy them lunch at the food court. His mother refused. "I will not eat in public. People are watching us."

Food was a constant struggle. Marielle tried to make Thai shrimp curry.

His mother looked at the shrimp and recoiled. "I will not eat rats!"

"Mom! They aren't rats."

"They are *Ighardayn!*" she insisted. *Mice.* She ate only bread and olive oil that night.

While Abdou was at work, a coworker rushed in. "Check the news. King Hassan II of Morocco died."

Abdou called home. "Dad, the King died."

"Don't say that, son! We will get in trouble."

"No, Dad. It's on the news."

"Was it a coup?"

"No. Illness."

The King's death cast a shadow over the trip, but his father returned to Morocco with a new perspective.

Back in the village, Abdou's father opened a small convenience store. He became a local storyteller, gathering old men around him to talk about America.

One day, a friend in the Moroccan security services called Abdou.

"Tell Uncle Mohamed to stop, or he will get in trouble."

"What did Dad do?"

"He keeps lecturing people about democracy in the US. He complains about his pension. He tells everyone he will resist the government because he receives 'foreign aid'—your remittances."

Abdou called his father. "Dad, they are watching you. Stop the speeches."

"Son," his father replied, defiant. "We have a new King now. Things are changing. I am too old to be afraid. Let them watch."

The next time Marielle visited, she brought him a peace offering: a suitcase full of Pringles, and ice cream in dry ice—the only American things he loved without reservation.

Abdou found that a lot of positive changes were taking place in Morocco, under the new king Mohamed the 6th. However, corruption remained rampant.

CHAPTER 26: The Dotcom Death March

One day, while visiting Morocco, Abdou sat with his cousin Ali in a coffee shop. One of Ali's friends leaned in.

"I heard you are a software engineer," the friend said. "I am as well."

"Nice to meet you. What kind?" Abdou asked.

"I am a WordPerfect engineer. Kind of. I graduate next month."

"From which school?"

"The School of word processing."

Abdou realized the young man was in a three-month typing course.

"To check that you are a real engineer, what does the F6 key do?" Ali's friend asked.

"I don't know," Abdou scoffed. "It depends on the application. It depends on how you programmed it."

Later, the young man swore to Ali that Abdou was a fraud, since everyone knew that F6 made text bold in WordPerfect. From then on, Abdou and Ali nicknamed him "F6."

The nick name stuck and everyone started calling the young man F6.

When Abdou returned to Minnesota, he found his financial portfolio decimated. The Dotcom bubble had burst. His high-tech stocks—Sun Microsystems, SAVVIS—were worthless. It was a devastating blow, but a valuable lesson.

Many of his former teammates had jumped ship to XTRCorp, a data storage startup. Lured by the promise of stock options and cutting-edge clustered data storage, Abdou joined them.

His old boss, TL, and his mentor, "Ryk," worked there. It felt like a reunion.

But within days, Abdou realized the product was a disaster. The software was riddled with hundreds of bugs. The hardware was unstable. Customers were returning boxes by the pallet load. Millions of dollars were being written off every quarter.

Abdou was on a "death march." He worked from 6:00 a.m. to 7:30 p.m., seven days a week. Alongside his ex-Marine friend Jeff, he fixed twenty bugs a week, only to see twenty more appear like hydra heads. The Java-based configuration software was a memory hog, causing storage nodes to deadlock and vanish from the data storage cluster.

Worse, there was a culture war. The "Old Guard" managers clashed with the incoming engineers. Whenever Abdou filed a critical bug report, a manager would downgrade it to "non-critical" to make the metrics look better.

"You can't put lipstick on a pig!" one engineer wrote in a bug report. "Do this, and the data gets corrupted!"

The manager marked that bug as "low priority" as well.

Desperate for a win, a board member forced the company to sign a contract with WebReplica, an Israeli data storage firm. The engineers were never consulted.

Abdou was assigned to integrate the software. Within two days, he found massive security holes. Anyone could read and write to the storage appliance through WebReplica's software.

He notified Ryk, the VP of Engineering.

By 3:00 p.m., Ryk called him in. "Hey, 'The Body,' pack your bags. We're flying to Tampa to meet WebReplica. Flight leaves at 6:45."

Abdou didn't have time to pack clothes; he only had his laptop. He missed lunch and dinner. That night at the hotel, the kitchen was closed. He survived on twelve airline peanuts.

The next morning, starving, Abdou hit the breakfast buffet like a snake unhinging its jaw. He ate two bowls of fruit, two pancakes, three hard-boiled eggs, and a yogurt.

The 9:00 a.m. meeting was tense. The WebReplica CEO, a former Israeli Defense Forces officer with a crew cut, glared at them.

Abdou presented his findings. A young female engineer on their team confirmed some of the vulnerabilities.

The CEO exploded. "We have a contract! You have to integrate our software as is!"

"We will revisit this once you fix the vulnerabilities," Ryk said calmly.

"No! I will call your board member!"

The meeting ended abruptly. Abdou, Ryk and Bob flew back to Minnesota.

Abdou wrote a detailed report, marking every bug as a "show-stopper." XTRCorp abandoned the integration, eating hundreds of thousands of dollars in losses.

The recession deepened. XTRCorp announced layoffs. The workload for the survivors doubled.

Ryk tapped Abdou to join a "skunkworks" project—a secret mission unauthorized by the board. Codenamed **Eclipse**, the goal was to build a new clustered file system using Linux and open-source tools.

When the Eclipse engineers unveiled their prototype, the Old Guard revolted. The resistance to change was fatal.

On April 11, 2002, Abdou got a call from his boss, JR. "Come to the office, bud. Something terrible happened."

JR was a terrific manager and engineer. He always called Abdou "bud."

The board had fired Ryk and the CEO. No explanation.

That night, Abdou met his friends SA and AK at Applebee's. They were devastated. Ryk was a great leader, and the company had just cut off its own head.

A week later, chaos erupted.

Someone exploited the very vulnerabilities Abdou had warned about. They hacked the payroll system and printed hundreds of copies of every employee's salary and stock options to every printer in the building.

A young engineer marched up to Abdou. "It is unfair that you get paid three times what I make."

"How do you know that?"

"Everyone has a copy. Go look at the printers."

Abdou sighed. "How many lines of working code do you write a week? Do I rewrite your broken code all day? If anything, I should add your salary to mine."

Abdou checked the printouts. Ironically, he found he was paid less than some of the "deadwood" employees.

The accountant was blamed for using an insecure storage device, but the manager who had downgraded the security bug to "low priority" went unpunished. Abdou kept a copy of the original bug report as insurance.

The new CEO called an all-hands meeting. It lasted five minutes.

"Everyone works on the old product. Eclipse is dead. If you aren't happy, resign by the end of the day. We will give you two months' severance."

By 5:00 p.m., most of the engineering team resigned.

Colonel E., the new VP of Engineering, called Abdou. "Don't resign. I need you."

Abdou negotiated hard. "I will stay if: I get one year of severance if fired. I keep my intellectual property. You remove the non-compete clause. And I want a 25% raise."

Colonel E. agreed to everything in writing.

Abdou couldn't resign anyway—Marielle had quit her job to raise their kids, and he needed cash to weather the storm.

He was promoted to replace JR. But without the great JR shielding him, Abdou clashed constantly with the managers who kept downgrading bugs.

"This product is unfit for the market!" Abdou shouted during one meeting. "We had a major data leak because of you!"

The writing was on the wall. Investors sued to recover their $25 million. The stock tanked.

Abdou, SA, AK, and TL started meeting in secret. They registered a new company, **Consolidated Data Systems**, to build software that would compete directly with WebReplica.

Marielle got a job at the University of Minnesota to secure health insurance for the family.

In August 2002, XTRCorp filed for bankruptcy.

When the assets were put up for sale, Abdou found his name on the list of engineers to be transferred to the acquiring company. He didn't want to go—he wanted his severance to fund his startup.

He went to Colonel E. "Lay me off. Keep Jeff instead."

"You are crazy. Why?"

Abdou handed him a copy of the original bug database, proving the management's negligence in the data leak. "I am worried about how the due diligence will go if I talk to the buyers."

Colonel E got the message. Abdou was laid off. Jeff kept his job.

But there was a final twist. The investors sued to freeze all employees and executives severance packages, including Abdou's.

Devastated, Abdou dipped into his savings. The Iraq War was starting, adding global tension to his personal crisis.

Months later, he finally received 90% of his severance. It was enough. Abdou was free to build his own future.

CHAPTER 27: September 11

Ten months into his job at XTRCorp, the world stopped.

On September 11, 2001, terrorists flew airplanes into the World Trade Center, the Pentagon, and a field in Pennsylvania, killing 2,977 innocent people.

Abdou watched the news in horror, his thoughts racing immediately to his former colleagues at Bridge Information Systems. They worked on the 92nd and 94th floors of the Twin Towers.

He spent the night frantically sending emails and making calls. No one answered. The fear of those hours left a lasting mark on his soul. Eventually, word came through: the New York team had evacuated safely. Abdou wept with relief.

But the world outside had changed.

On September 12, XTRCorp held a mass layoff. The atmosphere was grim.

The next day, Abdou stepped out of the office building for fresh air. He saw a circle of engineers in the parking lot. Some had been laid off the day before and had returned to collect their belongings.

As Abdou walked closer to say hello, he froze. One of the men was holding a Kalashnikov rifle.

Panic surged through him. *Was this a workplace shooting? A follow-up attack?* Abdou spun around and sprinted back into the office.

There was no attack. It turned out the engineer was moving to California to try his luck in Silicon Valley and was giving his gun collection to a friend before biking across the country. But in the shadow of 9/11, Abdou's innocence was gone; he now imagined the worst.

The fear followed him inside. During a coffee break, he joined a group of engineers discussing the attacks.

"The Muslims did this," one engineer spat.

"We should kill all Muslims," another agreed.

Abdou looked at the man speaking. It was a "loudmouth" colleague whom Abdou had helped countless times. He had fixed his bugs, saved his code, and covered for his mistakes.

Fear turned into cold anger.

Abdou walked into the office kitchen. He grabbed a large knife used for cutting birthday cakes. He walked back to the group and held the handle out to the loudmouth.

"Here you go," Abdou said, his voice shaking. "I am a Muslim. Start with me."

The group went dead silent. The loudmouth turned pale.

"No! Not you! Not you, Abdou," he stammered. "I am talking about the *others*."

"So, you want to kill every other Muslim except me?" Abdou asked.

The group dispersed, heads down.

Abdou felt sick to his stomach. He told his boss he was going to see a doctor, then went home. He didn't return for three days.

That night, a knock came at his front door. Abdou froze. Was it the backlash?

It was Jerre, his neighbor and friend.

"Let's have a beer at my house," Jerre said. "I need to talk to you."

Abdou walked next door. Jerre pointed to a high-powered rifle leaning by his front door.

"I have your back," Jerre said firmly. "There are a lot of assholes and bigots looking for scapegoats right now. I will keep an eye on your family, my dear neighbor."

Abdou was touched by the gesture, though the sight of the gun was a stark reminder of the danger.

It was a dark time. Attacks against Muslims were rising. Even members of the Sikh community were being attacked by ignorant bigots who confused them with Muslims. Abdou stopped going out, except to buy food. He lived in fear of an inevitable backlash.

Years later, when Jerre tragically drowned while scuba diving, Abdou mourned him deeply. In the darkest week of Abdou's life in America, Jerre had been the one to stand guard.

CHAPTER 28: Consolidated Data Systems Startup

Consolidated Data Systems's "headquarters" was a closet inside a carpet store in Hopkins, Minnesota.

The office consisted of two tiny rooms totaling less than 100 square feet. Abdou and SA turned one room into a lab, building workbenches from plywood and buying cheap IKEA chairs. The other room served as a server closet and conference room.

It was cramped, hot, and smelled faintly of wool. The engineers sat knee-to-knee. Nerves frayed, and tempers flared, yet they collaborated furiously to release the first version of their software.

They operated on a shoestring budget. Abdou, SA, AK, and TL flew to Philadelphia for the USENIX LISA conference—their first trade show. It was there that they met engineers from AOL who saw a demo of their software.

Back in Hopkins, the stress was palpable. No one was drawing a paycheck. They were running on fumes until an investor from Japan, Mori-San, stepped in with $275,000.

It was a lifeline, but it came with a catch. Mori-San demanded exclusive distribution rights in Japan. He never sold a single copy, effectively locking Consolidated Data Systems out of the entire Japanese market. But his cash kept the lights on.

Abdou enrolled in Saturday courses at the University of St. Thomas to sharpen his business skills. He also studied Japanese culture to better understand Mori-San.

When the investor visited Minnesota, he was impressed by the operation's frugality. Seeing brilliant engineers crammed into a carpet store backroom convinced him that his money was being used efficiently.

Abdou invited Mori-San, SA, and AK to his house for a Moroccan couscous dinner. He remembered a lesson from his cultural studies: *Never let a guest's cup run dry.*

Every time Mori-San took a sip of wine, Abdou immediately topped off his glass.

"Abdou," AK whispered, concerned. "You are trying to intoxicate the man."

"No, it is respect," Abdou whispered back.

Mori-San enjoyed the hospitality (and the wine). Fortunately, AK, who didn't drink, drove the investor safely back to his hotel.

With the new funding, Consolidated Data Systems signed a large telecom customer and moved out of the carpet store into a real office in Hopkins. Everyone finally had a desk. However, the expansion accelerated their cash burn rate.

Then came the big win.

In early 2005, NASA contracted Consolidated Data Systems to supply software for the Space Shuttle program. Following the tragic explosion of the Space Shuttle *Columbia* during re-entry in 2003, NASA needed better real-time data monitoring. Consolidated Data Systems's software allowed scientists at three different space centers to monitor the shuttle data simultaneously during takeoff and re-entry.

Despite the prestige of the NASA, DOJ, and DOD contracts, the company's revenue remained lumpy. By mid-2005, they were on the brink of failure again.

SA worked his magic and secured a critical $100,000 investment from the University of St. Thomas. It saved the company.

Then, a ghost from the past returned.

AOL contacted them to negotiate a site license. Their engineers had seen the demo in Philadelphia three years earlier and had tried to reverse-engineer the software to avoid paying for it. They failed.

Now, they were back, checkbook in hand.

CHAPTER 29: Outsourced Freedom

Thanks to SA's strategic maneuvering, the founders leveraged the pending AOL contract to sell Consolidated Data Systems. However, the company's valuation took a hit following an internal incident involving a specific team member.

Although Abdou and SA often butted heads, they formed a united front to navigate the sale.

The legal process was messy. Some sneaky lawyers tried to insert language into the public contract disclosing that *"A member of Consolidated Data Systems was..."* and then providing minor details of the incident. Abdou and SA refused. If the text remained vague, every team member would be a suspect, staining the reputations of the innocent.

They fought back. Finally, the lawyers agreed to redact the specific details. The final sales contract filed with the US Securities and Exchange Commission (SEC) read:

> "* Certain confidential information in this document has been omitted pursuant to a request for Confidential Treatment and filed separately with the Commission."*

It cleared Abdou and SA of any wrongdoing, but the experience shattered Abdou's trust in corporate politics.

Despite the drama, the sale earned the founders a significant amount of money. They joined the acquiring company for a mandatory one-year transition period, during which they signed strict non-compete agreements.

One afternoon, while Abdou was working in his office in Hopkins, an employee transferred a call. "Someone from the Moroccan embassy is on the line."

Abdou picked up. "Hello?"

"Hello, Abdou. This is Reda from the embassy."

"Hello, Reda. I hope you are doing well."

"It was very easy to find you," Reda said pleasantly. "Your name and phone number are all over the internet. I am calling about a post you made in support of *Telquel* magazine. As you know, Morocco has changed a lot."

Abdou felt a chill. *Telquel* was an independent Moroccan magazine known for its criticism of the government. He was being watched.

"Yes, it has, Reda," Abdou replied, keeping his voice steady. "So what is the problem? I am for a free press. I am against the government's suppression of it."

"We are proud of what you've achieved," Reda pivoted smoothly, ignoring the challenge. "I am the Moroccan embassy's economic adviser in the US. I want to see if you can do something for Morocco."

"Yes. I am thinking about doing that exact thing. I am already involved in the financing of a preschool."

The pressure continued to mount. A few months into the transition, disaster struck—not in the boardroom, but in the operating room.

During a routine surgery, a medical mistake caused a massive staph infection. Abdou was hospitalized for over a month. For twenty-eight days, he could not eat solid food. His fever spiked and lingered for weeks.

It was a nightmare for Marielle, who was six months pregnant with their daughter. She worked at the same hospital, which turned out to be Abdou's saving grace. She had him sign authorizations that allowed her to track his charts in real time.

One day, a nurse entered the room with a syringe. She was about to inject him with insulin.

"Stop!" Marielle ordered. "He is not diabetic!"

The nurse had the wrong patient file. If Marielle hadn't intervened, Abdou could have gone into shock and died.

Even in his hospital bed, weak and starving, the phone rang.

"I heard you are sick," Reda said. "The Ambassador and I are in Minneapolis meeting with expats. He is leaving for DC, but I am coming to visit you in the hospital."

Reda visited, polite and concerned. They spoke frequently after that, until one day Reda called to say goodbye. He claimed he was joining the Ministry of Foreign Affairs in Rabat. Abdou tried to locate him for years afterward but never found a trace. He suspected Reda had vanished into the intelligence services.

Abdou couldn't stop working. Between fevers, he opened his laptop on his hospital tray and fixed bugs and implemented software enhancements.

A new engineer from the acquiring company began emailing him with questions about the code. Eventually, they moved to Skype.

As Abdou walked the new engineer through the architecture, the man mentioned his team in India.

Abdou froze. He realized the truth: The company had hired an offshore team. He wasn't just helping a colleague; he was training his replacement. His position had already been outsourced.

When Abdou was finally discharged from the hospital, he was done. He had negotiated forty-five days of vacation and sick time in his contract.

He took it all.

He went to visit his parents in Morocco. He met with university presidents and deans, and two engineering schools asked him to advise students on their final thesis projects. He was finally doing what he told Reda he would do—helping Morocco on his own terms.

When Abdou returned to Minnesota in November, he found a letter in his mailbox.

Notice of Termination.

He smiled. He was relieved.

The next day, he founded a new company based in Eden Prairie, Minnesota. He was free.

CHAPTER 30: Abdou's New Startup

Abdou used his newfound freedom to build a bridge.

He established a scholarship program to bring fifth-year engineering students from Morocco to the US. Although his non-compete clause bound him for another year, he was happy. He swore he would never work for another boss again. He shuttled between Minnesota and Morocco, spending time with his family and bringing students to Minnesota to collaborate on research.

In 2010, Abdou decided to return to school for a third time. He enrolled in the Executive MBA program at the University of St. Thomas.

There were bright spots. He reconnected with his old boss, JR, and met wonderful people like Kakie. Most of the faculty were kind and thoughtful teachers.

However, the social climate was hostile. It was the height of the Tea Party movement, and many students in his cohort wore their politics on their sleeves. They didn't want him there.

The university held a "Winter Term Residency" in January, where students stayed at a hotel near the Minneapolis campus to foster cohesion. They ate lunch and dinner together and gathered for happy hour after class.

Instead of cohesion, Abdou found exclusion. He knew about the racist incidents that had occurred at St. Thomas after 9/11, but experiencing hate, himself, within the confines of higher education was a different kind of shock.

One night at happy hour, a female student turned to him. She was drunk, but her words were sharp.

"No one likes you here," she slurred. "This is a Catholic university. Why are you here among us?"

A management professor sitting at the table looked horrified. He intervened immediately, reprimanding the student.

Abdou was hurt, but he refused to retreat.

"I chose St. Thomas because it offers the flexibility to study and work at the same time," Abdou replied calmly. "I have every right to be here. Perhaps you will learn something from my presence."

Without the support of Kakie and the professor, Abdou might have dropped out the next day. Instead, he decided to stick it out, determined to shine and prove that he belonged.

That same December, Abdou's mother and his sister, Kadouj, arrived in Minnesota. They had come to help his other sister, Fatoum, who was about to give birth to her son.

The Minnesota winter was a shock to their systems. They were used to Moroccan winters, where lows dipped to 28 degrees, and highs reached 60 degrees. Minnesota was a frozen locker.

Abdou took them to Fairview-University of Minnesota Hospital to visit Fatoum.

His mother and Kadouj were stunned. The room was spotless. The technology hummed quietly.

"Did you bribe the nurses or the doctors to get this room?" his mother whispered.

"No, Mom," Abdou smiled.

"I don't see cats roaming the hospital like in Morocco," Kadouj observed, looking down the hallway. "And the nurses... they aren't yelling at the patients. They aren't yelling at the visitors."

"No," Abdou replied. "They don't do that here."

The joy of the new baby was cut short by a phone call from Morocco. Abdou's brother was on the line.

"Dad had a stroke."

Panic set in. Abdou had to get his mother home immediately. But he couldn't go with her—he had school and work. Fatoum had a newborn. Kadouj had to stay to help.

It was a terrible dilemma. His mother was illiterate; she spoke neither English nor French. Even her Arabic was not good. She only spoke Tamazight.

Abdou changed her ticket. He wrote a series of notes in English and French, just as he had done for his parents' first visit: *"I do not speak English. Please help me find my gate."*

He handed her the notes and hugged her goodbye at the security checkpoint, praying she would be okay.

Against the odds, navigating through international terminals with only folded pieces of paper, she made it home to her husband.

CHAPTER 31: Goodbye Mother

Abdou received the call he had dreaded for years. It was his brother-in-law.

"Salam, Abdou. Unfortunately, I have bad news. Your mother died this afternoon."

Abdou could not speak. He could not cry. He was simply numb. The shock paralyzed him.

He was heartbroken. He hadn't seen his mother in two years.

He flew to Morocco for the funeral. A member of the intelligence services, a contact of his brother, met him on the tarmac and whisked him through border control and customs, bypassing the lines.

His brother's driver was waiting. As they drove toward the village, the driver stopped at a gas station.

"Titi, your sister-in-law, sent you some dinner," the driver said, handing him a container.

Abdou ate in silence. Titi was one of the kindest people Abdou had ever met; everyone in the family, including Marielle, loved her like a sister.

They resumed the trip, the car cutting through the dark night.

Abdou arrived in the village at 2:30 a.m.

It was too late. In accordance with Islamic tradition, Abdou's mother had been buried two days before he arrived. He stood in the house she once filled, heartbroken that he hadn't been there to say goodbye.

In the morning, Abdou awoke to a house full of movement. Several groups of women he didn't recognize were sweeping the floors and doing laundry. As the day went on, one group left, and another arrived to cook for the continuous stream of visitors.

Women he did not know washed his clothes, folded them, and placed them on his bed.

Abdou realized this was the old village solidarity in action. Even though he had been away for decades, the community still held him and his family in its heart. He was profoundly thankful for the tradition. For weeks, people continued to bring food, or cook at home, so the grieving family wouldn't have to lift a finger.

In the afternoon, Abdou went to the cemetery.

His sister led him to his mother's fresh grave. He stood over the mound of earth, praying and paying his respects. He then visited the graves of his other elders buried nearby.

As he walked through the cemetery, he noticed broken headstones and trash. Vandals had desecrated some of the graves. Abdou was furious. That such disrespect could happen in a Muslim country, in a place of rest, made his blood boil.

When he returned home, he found his father in the living room with an Imam. The Imam was reciting verses from the Quran. His father read along, tears streaming from his deep, gazing eyes.

Abdou was touched by his father's vulnerability. He approached the Imam, handed him 500 Dirhams as a donation, and left the room.

Outside, he learned that the Imam had already collected 4,000 Dirhams from his brothers and other visitors. The anger returned. Abdou felt the man was exploiting the sad event, using a grieving family as a cash machine.

He stayed for the mourning period, soaking in the village solidarity and the sorrow of the empty house.

A few days later, Abdou returned to America. He had to pack. His internship in Asia was about to begin.

CHAPTER 32: The Global Systems

Abdou was assigned to a three-person study group for the class trip to Singapore and Malaysia. His partner, TW, was a national intelligence officer. Their task was to assess the risks of industrial espionage for a Minnesota company and to compare the two nations' global systems.

In Singapore, Abdou was struck by the country's flawless efficiency. He argued that strict order and control were essential for a developing nation to join the ranks of the First World—a view that put him in direct conflict with some of his American classmates, who prioritized abstract liberty over functional development.

The team received high-level briefings from CIA analysts, U.S. Embassy economists, and military officers. Abdou asked the officers about the "Pivot to Asia"—the new doctrine that shifted the U.S. military's focus to counter China's growing influence.

A few of his arrogant classmates groaned. They had no idea what he was talking about; their worldview was limited to domestic cable news, while Abdou consumed international analysis in four languages.

The cultural disconnect turned dangerous in Malaysia.

The student cohort was invited to dinner at a large Islamic center. As a gesture of goodwill, the organizers gifted every student and faculty member a bag containing pamphlets and a copy of the Quran.

During dinner, Abdou overheard a student whisper, "Let's burn the Quran."

Abdou froze. The timing couldn't have been worse. Just a week earlier, U.S. troops had burned Qurans at the Bagram Air Base in Afghanistan, sparking riots that killed over forty people. Tensions were razor-thin.

After dinner, Abdou quietly approached the hotel manager.

"I have a grave concern," Abdou said. "Some of my classmates may plan to desecrate the Quran copies in their rooms."

"What are you saying?" the manager asked, confused.

"Please," Abdou urged. "Go to every room the moment our group leaves in the morning. Check the bathrooms. Do it before your cleaning crew finds them. If your staff finds desecrated holy books, there will be violence. I do not want a riot here like in Afghanistan."

"Oh my God," the manager whispered, paling. "What kind of people are these?"

"Students from a prestigious Catholic university, sir."

The next day, the manager confirmed Abdou's fears. He had found damaged copies of the Quran in the garbage and bathrooms. Abdou had prevented a diplomatic incident, but he was disgusted.

He just wanted to graduate and escape.

In his final class, the conflict with an adjunct faculty member came to a head. Abdou received a "B" on a group project report, while his three teammates received "A"s. This was inexplicable—Abdou had written the entire report himself because his teammates were busy with work and Abdou had flexibility.

He emailed the professor: *Why did I receive a lower grade on the report I wrote, while the other members received A? It was the same report?*

The professor's reply was shocking: *"Is your mother Asian? You still got a B."*

Abdou was furious. The professor was relying on a racist "Tiger Mom" stereotype, assuming Abdou was only complaining because of cultural pressure to be perfect.

Abdou fired back: *"My mother is not Asian. And even if she were, she died recently. My pain is still raw. What you said is inappropriate and racist."*

Abdou filed a formal complaint. He does not know if the professor was ever reprimanded for his behavior.

It was fitting that this was the same professor who had told Abdou in 2012 that he was in the wrong business. *"Cloud computing is a terrible idea,"* the professor had scoffed. *"It will collapse."*

Abdou ignored him. He continued to build his Cloud and AI business. Years later, when Abdou signed the Social Security Administration and the U.S. Department of Defense as customers, he was tempted to send copies of the checks to the professor.

Today, Cloud Computing is a trillion-dollar industry. The professor was wrong about the future, and he was wrong about Abdou.

Abdou graduated at the top of his class in 2013 and was inducted into the Beta Gamma Sigma international honor society.

He continued to build bridges back home, advising students at Moroccan engineering schools. He mentored Khalid, a PhD student who faced harassment from a new dean at his university. Abdou fought a bureaucratic war to protect his protégé. Despite the obstacles, Khalid earned his PhD in machine learning with high distinction and became a professor himself.

CHAPTER 33: Goodbye Father

Abdou flew to Morocco to visit his family. He was impressed by the changes he saw in the big cities, but the rural areas remained almost exactly as he had left them, save for wider coverage of electricity and running water—even if the water was intermittent. Poverty was still rampant.

He found his father getting old, but he still retained his good humor. One afternoon, a former classmate of Abdou's came to visit.

"I want to talk to you about something. Let's step outside," the friend said.

Once they were out of earshot, the friend lowered his voice. "I saw your father several times sitting alone under the olive tree. He goes there once in a while and he cries. I asked him if he was okay, but he never replies. Is he okay?"

Abdou felt a weight settle in his chest. "Thanks for telling me. I did not know any of this. I will try to find out."

Later that week, Abdou and his sister took their father to the city to do some shopping. Abdou knew his dad loved black coffee, so he steered them toward a café.

"Dad, want some coffee?" Abdou asked.

His sister jumped in right away. "No, he cannot have coffee. His health does not allow it. Get him an avocado juice."

Abdou's father's face fell, looking like a child who had been scolded. But the moment Abdou's sister left to shop for food, Abdou signaled the waiter and ordered a cup of espresso.

"I really miss coffee, son. They control what I eat and what I drink," his dad whispered, gulping the hot espresso in one shot and looking around to make sure his daughter didn't catch him. He smiled, a spark returning to his eyes. "Thank you, son. Thank you."

The next day, Abdou took him to another coffee shop in the nearby town of Afourer. This time, he let him have two small cups. But while talking to him, Abdou realized that his father's mind was drifting. He had started to forget things. He even struggled to recall the names of his own granddaughters.

On the day Abdou was leaving the village to go to Rabat, he tried to give some money to his father. His father pushed it away, telling him to give it to his sister instead. He was afraid he would just misplace it.

Then, the old man started to cry.

"Your mother raised you well, son. May God save her soul. May Allah protect you. Please say hello to Marielle, your children, and Marielle's family."

As Abdou drove to Rabat, he wept, the image of his frail father burned into his mind.

A week after he got back to the U.S., on April 8th, 2017, he received bad news: his father had suffered a massive stroke. Abdou was devastated. He managed to go back one last time to see his father, who remained bedridden for almost a year before he passed.

When the end finally came, Abdou could not leave for the funeral; by custom, his father was buried within twenty-four hours.

Abdou faced a terrible dilemma of the heart. His father had died, yet life in America was blooming. His son, Salah, had just graduated from medical school; his daughter, Malika, had graduated from college; and his son was getting married. He was torn between the grieving and the celebrations, trapped between the sorrow of the past and the joy of the future.

With his parents gone, and the lingering pain too sharp to touch, Abdou did not visit Morocco for five years—breaking a lifetime habit of returning home every year or two.

CHAPTER 34: The End

Abdou had finally achieved the financial and professional autonomy he sought since leaving Morocco. The chaos of the dot-com era, the fear of 9/11, and the humiliation of the porter job felt like distant echoes. Advising students and traveling often, he settled into a satisfying rhythm that left him time for family and professional passions.

His ongoing work included mentoring young engineers at the Moroccan engineering schools he had once attended, a small act of repaying the kindness and patronage he had received years earlier from figures like Petit and Troeh. This work, coupled with Marielle's enduring dedication to research, anchored their lives not just in success but in meaning.

Yet the ultimate fulfillment of his sacrifice was written not in a contract or a severance package, but in the lives of his children. Abdou and Marielle watched their eldest complete their university careers, their diplomas the tangible proof of the opportunity he had chased across the ocean.

But this triumph was always shadowed by the shifting political winds of their adopted home. Despite their citizenship, the rising tide of anti-immigrant sentiment, the fear of ICE raids, and the normalization of xenophobic rhetoric created a constant, low-grade anxiety for Abdou, Marielle, and their children. Their hard-won security, paid for in blood and tears, remained conditional, vulnerable to the latest political upheaval. This persistent fear was the ultimate, enduring cost of their success.

The commencement day was everything Abdou had dreamed of, a moment where the full weight of the past 30 years culminated in blinding joy. He stood proudly among the cheering crowds, feeling a connection to the village elders and his own father, who had preached the value of hard work and rising up after a fall.

His son, who had completed his MD and the brutal years of residency, had returned to Minnesota as a specialist in Trauma, Acute Care Surgery, and Surgical Critical Care. With expertise in advanced life support like ECMO, he dedicated his life to healing the most critically ill. His daughter launched a vital career as a GIS developer in solar energy, mapping the future of sustainable power. And his youngest, following the path of both knowledge and movement, began her studies at the University of Minnesota.

Abdou looked at his children, successful, secure, and thriving. He felt the phantom aches of the Lake Street apartment, the crushing weight of the $12.20 paycheck, and the deep, permanent sorrow of not being able to say goodbye to his mother. The costs were immense, and even in this moment of absolute generational triumph, the fight for their dignity continued. The circle of exile was closed, but the battle for acceptance remains unfinished.

ACKNOWLEDGMENTS

I would like to thank Marie, Krystal Hanson, Dr. Mustapha Naimi, Dr. Said Ouattar, and Dr. Hilali Abderrahman for reading early drafts of this book and for giving me valuable feedback.
To my late parents, Fadma and Mohamed, who sacrificed so much to educate me.
To my late in-laws, Bill and Carole, who accepted me as one of their own.
To my wife, Marie, who encouraged me to finish this book and who verified and corrected me when I did not remember some events correctly.
To Salah, Malika, Kenza, Zack, and Michelle.
To Krystal Hanson, who encouraged me when it was hard to relive and write about some of these events.
To all the geeks like me.
To the ones who face hardship in their lives. To the voiceless, the oppressed, and the dispossessed. To the ones who strive to do good on earth.
To IAV Hassan II, which taught me to analyze, pay attention, and care for rural people. And to my sociology professor Paul Pascon and IAV's Abdallah Bekkali.
In memory of Saoud Boulmene, Madame Petit, and Ahmed Darmouch.
A. El Haddi
The Republic of Chaska

January 18th, 2025

ABOUT THE AUTHOR

Abderrahman A. El Haddi is a software engineer, entrepreneur, and bridge-builder. Born in a small village in the Atlas Mountains of Morocco, he went from guarding family goats to securing technology contracts with NASA and the U.S. Department of Defense.

A veteran of the U.S. tech industry, Abderrahman navigated the Dotcom boom and bust, eventually founding his own successful software companies, Consolidated Data Systems and EnduraData. He holds an **Executive MBA** from the University of St. Thomas, where he graduated at the top of his class. He also holds a **Master's in Computer Science** from the University of Minnesota and a **Master's in Agricultural Engineering** earned jointly from the University of Minnesota and IAV (Institut Agronomique et Vétérinaire Hassan II). Abderrahman is a senior member of the Electrical and Electronics Engineers (IEEE) and a member of the Association for Computing machinery (ACM).

Today, Abderrahman lives in Minnesota, where he continues to run his software business while mentoring the next generation of Moroccan engineers through scholarship programs and research partnerships. *The Data Shepherd* is his first book.

www.ingramcontent.com/pod-product-compliance
Lightning Source LLC
Chambersburg PA
CBHW020926090426
42736CB00010B/1058